L. I. T. Library

Direct Digital Control for Building HVAC Systems

Second Edition

Direct Digital Control for Building HVAC Systems

Second Edition

Michael J. Coffin
Vice President
PG&E Energy Services Corporation
San Francisco, CA

Kluwer Academic Publishers

Distributors for North, Central and South America:
Kluwer Academic Publishers
101 Philip Drive
Assinippi Park
Norwell, Massachusetts 02061 USA
Telephone (781) 871-6600
Fax (781) 871-6528
E-Mail <kluwer@wkap.com>

Distributors for all other countries:
Kluwer Academic Publishers Group
Distribution Centre
Post Office Box 322
3300 AH Dordrecht, THE NETHERLANDS
Telephone 31 78 6576 000
Fax 31 78 6576 254
E-Mail services@wkap.nl>

Electronic Services <http://www.wkap.nl>

Library of Congress Cataloging-in-Publication

Coffin, Michael J. (Michael James)
Direct digital control for building HVAC systems / Michael J. Coffin.--2nd ed.
p. cm.
Includes bibliographical references and index.
ISBN 0-412-14531-6 (hb : alk. Paper)
1. Commercial buildings--Heating and ventilation--Control.
2. Industrial buildings--Heating and ventilation--Control.
3. Commercial buildings--Air conditioning--Control. 4. Industrial buildings--
Air conditioning-Control. 5. Digital control systems.
I. Title
TH7392.C65C63 1998
697--dc21 97-46127
CIP

This printing is a digital duplication of the original edition.

Printed on acid-free paper.

Printed in Great Britain by IBT Global, London

Contents

Preface

Purpose of This Book

The purpose of this book is to introduce and explain direct digital control (DDC) systems used for controlling air conditioning systems in commercial and industrial buildings. Having now been in common use for more than a decade, DDC systems can be found in virtually every kind of building imaginable, managing environments from the critical to the commonplace. This text is a survey of DDC systems from the perspective of the end user and is intended to provide the reader with an understanding of the basic construction of these systems, as well as their benefits and limitations. Basic information on control systems is provided in Chapters 2 and 3, but the focus of this volume is on the architecture of DDC systems and how they are best used to manage occupant comfort and energy conservation. Although written for building industry professionals who have some understanding of the construction of HVAC systems, it is my intent to present this information for the benefit and understanding of anyone requiring knowledge of DDC systems. It is intended as a book "for the rest of us," to borrow the familiar slogan of Apple Computer, Inc.—the many who must specify, purchase, manage, and support microprocessor-based building control systems and who may have only a moderate knowledge of digital technology and how it works.

In the chapters that follow we examine basic control systems and computer fundamentals; how the components of these systems are brought together in unique ways to control HVAC systems; the media used to connect these systems to the equipment they control and to their human hosts; the languages they use to communicate between themselves and their human hosts; the technologies available to translate communication between systems of dissimilar manufacture; ways to use digital controls to manage building environments efficiently; how to reduce first cost through careful design and specification practices; methods for accurately estimating first cost; and what the future of DDC systems may hold. An equally important objective of this book is to provide readers with both a clear understanding of the many improvements these systems offer over conventional control methods and the knowledge needed to make informed decisions when approaching control system applications.

Unlike engineering textbooks, this book is written from an applications perspective; how these systems are used is given priority over the underlying theory. As such, from time to time terms are used that some may consider proprietary, based on my familiarity with certain products and systems. However, every attempt has been made to describe hardware, software, and system applications in the most universal terms possible. I have also avoided the use of technical language or complicated formulas when explaining concepts and applications. My hope is that the easier this book is to read and understand, the more useful it will be.

We begin with an explanation of what DDC is and how it differs from traditional control methods. If some of the terminology is unfamiliar, don't be discouraged. Chapters 2 and 3 review the fundamentals in preparation for more detailed discussions in later chapters.

Organization of This Book

Each chapter presents information sequentially, with subsequent chapters building on information previously presented. The reader should be able to proceed through the text without having to refer constantly to the index or glossary.

Chapters 1 through 3 address the fundamentals of control systems, and the reader should understand these thoroughly before continuing. How DDC systems connect to the equipment they control and how they communicate is presented in Chapters 4 and 5. Chapter 6 reviews proven ways to use DDC controls to achieve comfort and conservation objectives, while Chapters 7 through 9 consider the evaluation, selection, specification, and investment levels required for DDC systems. Finally, in Chapter 10 we will consider the future of DDC and the emerging technologies that could become future standards.

Chapter 1

Introduction to Direct Digital Control Systems

Definition of Direct Digital Control

Direct digital control (DDC) is a control process in which a microprocessor controller constantly updates an internal information database by monitoring information from a controlled environment and continuously produces corrective output commands in response to changing control conditions. This process is similar to the way traditional control systems work, except that digital information processing produces a much faster and more accurate control system response under all conditions at all times.

Going further, DDC is a three-step signal conditioning process in which:

1. Input information signals are converted from analog to digital format so that the microprocessor controller, which understands digital information, can read the incoming data.
2. Control commands are performed by the microprocessor incorporating this newly converted data.
3. The corrective output response of the controller to the device it is controlling is reconverted from binary data into an analog output signal that is recognized by the controlled device.

The process of routing information into and out of the digital controller is graphically represented in Figure 1-1. Information first enters the controller through a termination board. The digital controller can address each control point in the system through an interface or input/output (I/O) card. The signal next passes through an analog-to-digital converter, which transforms the control signal from an analog value to a binary value. Now the controller can interpret the meaning of the control signal and can act upon this information according to the instructions of the control programs in its memory. The output response of the controller is sent through a digital-to-analog converter, which returns the signal to its original analog format. This signal is then sent to a controlled device either directly as a digital signal or,

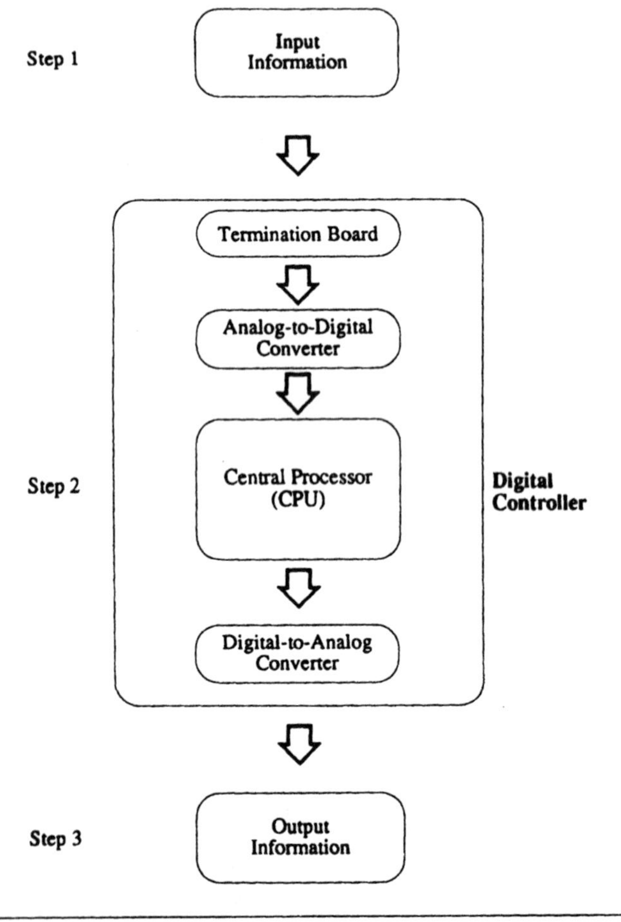

Figure 1-1 Information path: digital controller.

acting through a signal transducer, as a signal appropriate to the device under control.

The term *microprocessor* often gives people the impression that computer-based systems are complicated. In reality, microprocessors offer a simple and organized way to build almost any electronic system. In some respects, microprocessors mimic human thought behavior. They are made up of circuits or functions that are easily understood and readily available. All processing systems, including the human system, can be broken down into the functions of sense, decision, memory, and action. Following is a simple explanation of how the system works.

The sense function receives information and transmits it to the machine for interpretation and response. This input information can be data about surrounding environmental conditions, such as temperature, pressure, or humidity. Often this information is encoded and must be decoded as part of the interpretation process. Through sensing elements, the system can receive information from the equipment it is controlling or directly from humans. In a microprocessor system, the functional units that perform the sense function are called *inputs*. Inputs can be in the form of a binary, or on/off, state, or they can be analog (i.e., continuously variable).

The decision function is similar to the reasoning function of the human brain. It handles all the computations, logical operations, and operational decisions. These decisions take into account the inputs and the information stored in memory. The microprocessor performs the basic arithmetic and logic functions required to control HVAC equipment.

The microprocessor must remember what it is to do, what information is available for use, what it does, and the results of what it has done. It must also remember a number of rules that will be used in making decisions, performing arithmetic, and controlling the system. The microprocessor system performing this function is *memory*, analogous to the memory function of the human brain. It holds information on the step-by-step sequence of operations the microprocessor is to perform. The memory also stores the instructions and information, or data, that are used in the sequencing of operations. This sequence is known as a *program*.

The fourth function is the action function. Once a decision has been made by the microprocessor, the microprocessor carries out the decision by using the action units of the system. The action units allow the microprocessor system to control something external to the system or to communicate information to humans or to other machines. The action unit can turn on a supply fan, close a valve, energize a heating coil, or perform some other control operation. The action unit may be a device that displays information so that it can be communicated to humans. The devices that implement the action function are called the *outputs* or controlled devices of the system.

Direct Digital Control versus Conventional Control

Throughout this book traditional mechanical or electric controllers, such as receiver–controllers or electric balanced–bridge controllers, are referred to as *conventional controls*; they are also known as single-loop controls (SLCs) or discrete controls because they control only one sequence or loop in an entire control system. In contrast, and because they possess electronic intelligence, DDC systems can control many control sequences simultaneously. It is important to understand the differences between digital control and conventional control methods. Pneumatic and analog electronic controllers perform control functions based on mechanical or electric signals and can handle only one loop in a fixed manner. Digital controllers follow defined program instructions that are stored in memory and can manage many control loops simultaneously. Furthermore, a digital controller can be reprogrammed as needed for differing control strategies, without changes in hardware. DDC offers several additional advantages over conventional control methods. DDC systems provide very accurate and repeatable control of set points. Unlike that of mechanical controllers, the accuracy of digital controllers will not drift over time owing to lack of maintenance or to mechanical fatigue. The resulting offset from set point that this drift causes gradually decreases the performance of the control system.

The operating functions of the digital controller can be adjusted through software to allow "fine tuning" of the control system. This adjustment takes place through a human-to-machine interface, usually a computer keyboard and monitor screen. This ability to adjust complex control strategies offers building operators an economic advantage because all control functions are integrated into a single control environment; a minor change to a control strategy in a conventional control system may require the replacement or modification of costly hardware, whereas the same change in a digital system is accomplished through the existing software without impacting field devices. This capability is especially important in buildings where changing conditions can occur, such as shifts in floor plan layouts and tenant turnover.

Another significant difference between conventional and digital control methods lies in the ability of the system automatically to adapt itself to changing conditions in the controlled environment. Using software with adaptive control capabilities, a digital controller can self-adjust program variables in response to changing environmental conditions. Self-tuning PID (proportional plus integral plus derivative), artificial intelligence, and nonlinear expert control methods are examples of functions that differentiate microprocessor systems from conventional control systems. Figure 1-2 compares the conventional sensor-controller–controlled device, or single-loop control method, to the digital method of controlling multiple control loops from a single controller. These and other capabilities of digital systems are explored in the chapters that follow.

Conventional Control Loop Example

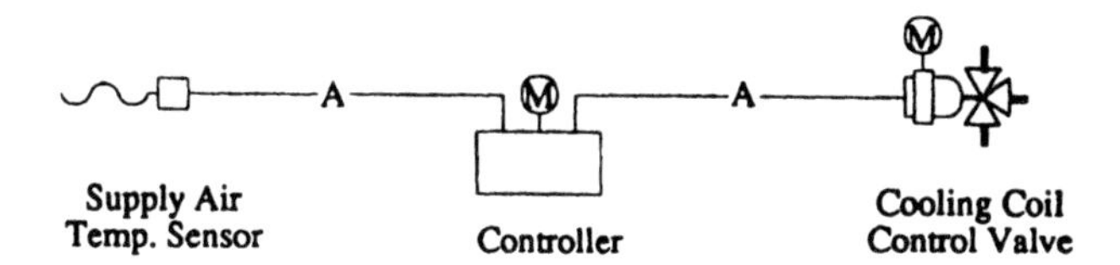

Digital Control Loops Example (Multiple Loops)

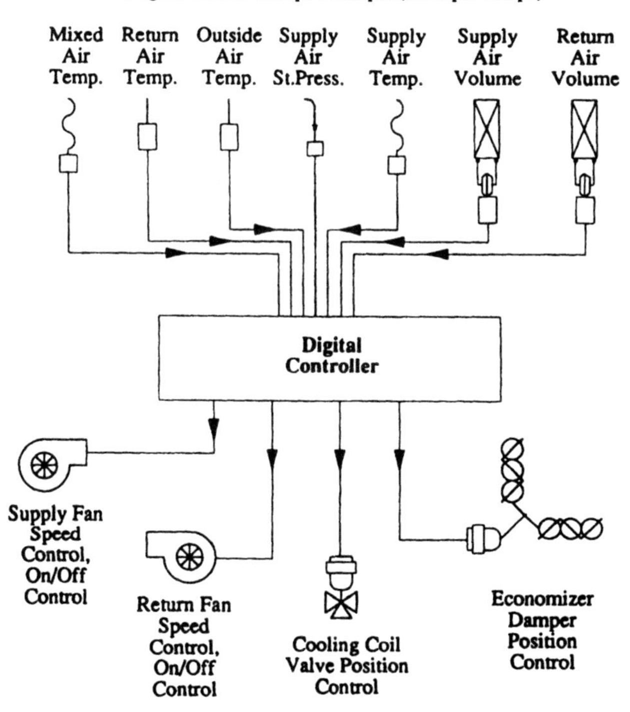

Figure 1-2 Conventional and digital control loop comparison.

Impact of Digital Controls on the Building Industry

There are many reasons why building industry professionals should understand the inner workings of DDC systems, especially now that microprocessor-based controllers have evolved into such a critical tool for managing human comfort, air quality, lighting, life safety, security, and energy consumption.

In our service-oriented society, our most precious commodity is information that is delivered in both an accurate and a timely manner. A DDC system, like any other microprocessor device, is a tool that is used to manipulate and process information. At its essense, it is an information appliance.

In their application to building HVAC systems, DDC systems provide building operators with valuable information that can be used to increase revenues and decrease operating costs in commercial buildings.

The largest operating expenses of a commercial building are electricity costs for lighting and for heating, ventilating, and air conditioning. Digital controls are being used to reduce energy costs, without creating an adverse impact on building occupants, through strategies such as load shedding, optimum start/stop, and reduction of the operating speed of electric motors. These and other energy-saving strategies are discussed in later chapters.

Digital systems are also making major contributions to the improvement of workplace environments. Commercial developers competing for building tenants must offer quality working environments that are comfortable, functional, and adaptable to the changing needs of growing businesses. Building developers are quick to recognize the value that an integrated building automation system can add to a property, as well as its value as a property management tool after a building becomes occupied.

Improving air quality is another opportunity for engineers to apply the capabilities of digital systems. Energy conservation strategies require that buildings be airtight to prevent the escape of costly heated or cooled air; this usually comes at the expense of ventilation. The "sick building syndrome" has become a national concern as researchers continue to uncover the effects of indoor air quality on human health and productivity in the workplace. Concerns over health hazards in the workplace and the spread of airborne contaminants are issues that have reached the forefront of public attention. The control of building ventilation is a problem that is being solved through the application of digital controls. In Chapter 6, ventilation control methods designed to improve indoor air quality are discussed.

In summary, effective and reliable building control systems are no longer an option; they are a necessity.

History of Electronic Controls

To appreciate the potential for DDC systems, it is helpful to understand the development of commercial control systems. Electric, pneumatic, and electronic control products are commonly used in commercial, industrial, and institutional facilities in new construction, reconstruction, and energy retrofit applications.

The first electronic control systems were introduced during the early 1950s. These systems combined vacuum-tube technology with resistance temperature-sensing elements. These temperature-sensing elements consisted of a variable resistor providing 1000 ohms of resistance at 70°F. The sensor provided an input to a vacuum-tube amplifier that positioned an actuator to control a damper or valve. During the late 1950s this system was refined further through the development of the control amplifier and bridge method.

The early 1960s brought major technological advances in the electronic component field. Control manufacturers began to use transistors, silicon-controlled rectifiers, integrated circuits, triacs, and other electronic components in the manufacture of commercial controls. These new technologies offered opportunities for increased product life and improved performance. In transistor-based control systems, the control amplifier and bridge assembly were enclosed in a single transistor operator. The operator could position valves and dampers to control not only control systems but remote devices as well. This led to the development of a combination electropneumatic (EP) transducer. This EP transducer is still used today in control systems for large buildings that require economical pneumatic actuation. Combining the cost effectiveness of pneumatic control devices with the accuracy of electronic controllers achieved a great improvement in overall control system performance over straight pneumatic control.

Later in the 1960s, transistor amplifier and bridge circuits were redesigned so that they could easily be placed on a wall to sense space conditions. The integration of temperature sensor and amplifier into a single enclosure became known as a sensor controller. This device made it possible to improve the control efficiency and to reduce the total cost by eliminating a device.

During the 1970s, control manufacturers began applying microprocessor technology to basic environmental control and to monitor and control energy consumption. Microprocessors coupled with pressure-sensitive keypads and light-emitting diode (LED) readouts formed a system that reduced building energy consumption by tracking building demand levels. By the late 1970s, control systems were offered that used central communications capabilities. This development made it possible to manage energy in several buildings from a single location through communication with control systems over conventional telephone wires.

In the 1980s, innovations that provided diagnostics and management reports were developed, to improve the usefulness of "smart" control systems. These developments brought improved operating performance, reduced installation costs, self-diagnostic routines, and improved techniques for digital communication to environmental control systems, and they are responsible for the explosive growth in this industry.

Development of Digital Control Systems

The advent of the microprocessor brought DDC techniques to the building industry. An early relative of digital controllers is the programmable logic controller (PLC), which has been used for many years for process control in manufacturing industries. PLCs are used in automobiles for engine control, dashboard instrumentation, and diagnostic checks during engine service and repair. Aerospace applications of computerized controllers are many, ranging from in-flight controllers to flight simulators. Modern communications systems rely on split-second digital processing of large amounts of information, which only a microprocessor can do.

The small size and relative lost cost of semiconductors have had a tremendous impact on the architecture of equipment used in the HVAC industry. The push for energy conservation measures resulting from rising energy costs during the 1970s introduced the need for control systems that could control building comfort and energy consumption. Two important needs surfaced when tighter control of HVAC energy usage became a concern: a controller that could perform more than one routine at a time, and feedback on system operation.

To perform complex energy-conserving control strategies, conventional control systems using single-purpose controllers had to be layered, resulting in hardware-intensive systems that were difficult to operate and service. This created the need for a multipurpose controller that could handle many control sequences simultaneously. Microprocessors are well suited for this purpose because they can perform multiple functions with great speed and accuracy.

When energy conservation strategies, such as variable air volume, became common in HVAC system design, the need for information on the status of different operating subsystems within the overall HVAC control system became necessary. Obviously, when building operators have many control routines running at the same time, it is difficult to gather information manually on the performance of the system; obtaining this information from a centralized point became a necessity. With this concept, individual control strategies could be monitored for their effectiveness in saving energy as well as providing useful information to the operator on the performance of the HVAC system.

The earliest form of hierarchical control system to develop from these needs was the centralized control and management system (CCMS) (see Figure 3-4). This system can integrate the control of the many control blocks in a total control system into a single system. Early CCM systems were designed around minicomputers, which were popular in business and scientific applications. They offered the power to process input data quickly and to send output control signals to controlled equipment from a central location. They also allowed an operator to monitor and modify control system performance from an operator's terminal, and they could display the results of system changes on a video monitor.

The configuration of these early machines was monolithic, meaning that the information-processing power of the system was housed in a large floor-standing cabinet next to a display monitor and keypad. These systems provided centralized processing of program routines and had no means to control HVAC equipment on an independent or local level. Input and output information was gathered in a local data acquisition panel (DAP) and then sent over the wires to the central processing unit (CPU) for information processing and decision making. The CPU then sent signals over wire to the devices being controlled. Although innovative for its time, there were many problems with this type of system. In systems with many inputs and outputs (known as points) the demands on the processing capabilities of the CPU would increase and the system would ultimately become bogged down and perform slowly. The microprocessors used in these machines were considered fast at the time but could not compute quickly enough to report changes in controlled variables instantaneously.

Because all data processing and control took place within a central computer, a system failure meant that the entire HVAC system would shut down or have to be operated in a manual mode. This created problems for building owners who were not sufficiently trained in computer repair and troubleshooting; by the time a service technician was called to help, many hours of productive time had been lost. The storage of control program routines and building trend data required data storage devices such as the external hard disk drives that held the software and program routines as well as recording alarms. These devices often could not store enough information to accommodate the data storage needs of a large system, which then required an operator periodically to transfer system data to floppy disks to create space on the drive. This configuration also required hardwire connections between the sensor input and output devices through the DAP and the connection of each DAP to the central system, resulting in installations that were wiring-intensive.

Developers of computerized control systems were on track with this concept, but it was not until the early 1980s that an explosion in the computer industry took place that would start the most significant period of change in HVAC control in over 90 years.

Perhaps the most significant impact on the use of computers in HVAC control systems came about after the introduction of the personal computer (PC) in 1981. International Business Machine (IBM) Corporation introduced the IBM PC, a microcomputer using an Intel 8086 microprocessor, a low-cost and reliable chip that has become the standard in the manufacture of intelligent control systems. Controls manufacturers soon recognized the opportunity to apply these systems in HVAC control applications and began to develop software to allow PCs to operate as host computers for hierarchical control systems.

Over the past several years, Intel has developed many enhancements to the original 8086 chip to increase its speed and capability, yet the state of the art of DDC systems remains based on 1981 technology and not the more recent versions of microprocessor chips, such as the 80286 and 80386. The reason for this time lag lies primarily in the conservative attitude of DDC product manufacturers toward technological developments that lack a track record. Early version chips are considered safe because they have performed reliably and consistently for many years and are in abundant supply. DDC systems often control critical environments where a system failure could affect human life and property; it is understandable that manufacturers faced with product liability tend to take a "wait and see" approach with new semiconductor platforms, even though they offer faster processing speed.

Furthermore, modern microprocessor platforms such as the Pentium® and MMX® offer far more capability than is needed to perform localized control of HVAC equipment. Microprocessor operating speeds of early chips are 4 to 8 MHz (megahertz), usually more than adequate to execute rapid control calculations in these control devices. As shown in later chapters, at the device level, where most HVAC control strategies are executed, a slower response often provides better management of the controlled variable.

At the central computer level, however, this is not the case. The faster a host computer can gather data from a network, process these data, and report changes in system operation to graphic user screens, the more valuable it becomes to a building operator. Because speed is beneficial at this level, the PC host systems being used today take full advantage of the most recent processor platforms.

Recent Developments in Direct Digital Control

It can be said that DDC systems have traveled a hard road to popularity since their introduction in the early 1980s. Early objections to the use of DDC centered around point density—that is, how control information is grouped and where it is located in the system—and the probability of system failure. Many of the early DDC systems were designed to control anywhere from 32 to 96 points in a single enclosure. Unfortunately , when a DDC controller failed, an entire control system was rendered inoperable. Because these first-generation controllers were so error-prone, the consulting engineering community approached the use of DDC with much caution and generally tended to favor the use of conventional control systems. To make matters worse, first-generation DDC controllers were difficult to program, often requiring knowledge of sophisticated programming languages on the part of building operators. This forced building operators to rely on manufacturers for ongoing support of their systems, and many operators complained because they could not make changes in control routines without hiring special programming assistance.

Fortunately, DDC systems have improved dramatically since these early systems were introduced. Advances in microprocessor technologies allow manufacturers to build systems with more control capability and better reliability.

The current trend in control system design is toward smaller, modular, stand-alone DDC panels. With microprocessor-based control at the local level (right at the HVAC equipment being controlled), the risk of a component failure is of minor consequence to the overall performance of the HVAC system. Furthermore, with the capability of supervisory computers to detect the failure of a control device and inform the building operator of the nature of the problem, it can be said that these systems are truly self-diagnostic and self-maintaining.

Acceptance of Digital Controls

In the past, design consultants resisted the use of DDC systems because of their relative newness and their high failure record in early product generations. In fact, what was deemed a failure in many cases was simply a misunderstanding of how the system was supposed to operate. As computer literacy has risen in recent years and as manufacturers have refined the capabilities of digital products, this resistance has significantly lessened. The main reason is that recent generations of microprocessors are much more reliable than earlier-generation products, and because silicon chips have been proven able to withstand severe ambient conditions and have longer life cycles than conventional control devices, thereby requiring less maintenance and calibration. Also, as innovations in silicon wafer technology continue to develop, the microchips that drive DDC products become less expensive, making digital systems available to cost-conscious builders and developers. As a result, DDC hardware is being introduced that is more compact, reliable, and affordable than ever before.

Innovations in the design and manufacture of DDC products have brought down the cost of these systems substantially, to the point where they can now compete on a first-cost basis with conventional pneumatic control systems. Respondents to a survey of energy management and control product manufacturers felt that pneumatic controls will lose their advantage of lower first cost to computer-based control products within 5 years. Eighty percent of the respondents thought that only certain pneumatic products will survive, such as pneumatic actuators for dampers and valves, because of their cost effectiveness in large applications.

Experts in the field of automatic control have questioned whether the additional capabilities that computerized controls can provide are really required for HVAC control applications. In a recent industry survey of controls manufacturers, control system users felt that 90% of the technology available in DDC products exceeded the abilities of those who are responsible for its implementation and use. The answer to this problem is education so that design, construction, and building management professionals can properly design and apply DDC systems.

A comparison between conventional and DDC systems, based on the criteria of system performance, first cost, reliability, flexibility, maintainability, ease of use, and life cycle cost, is shown in Figure 1-3. It is easy to recognize that DDCs can perform complex control sequences far beyond the capabilities of single-purpose controllers.

The development of commercial and industrial buildings has kept pace with the strong demands of service sector companies in the bull market 1980s. Aggressive competition between building developers for tenants has motivated developers to consider new and innovative ways to enhance their properties to attract tenants. Quality working environments that are comfortable, functional, and adaptable to the needs of the occupants have become the new standard. Building developers are quick to recognize that an integrated building automation system

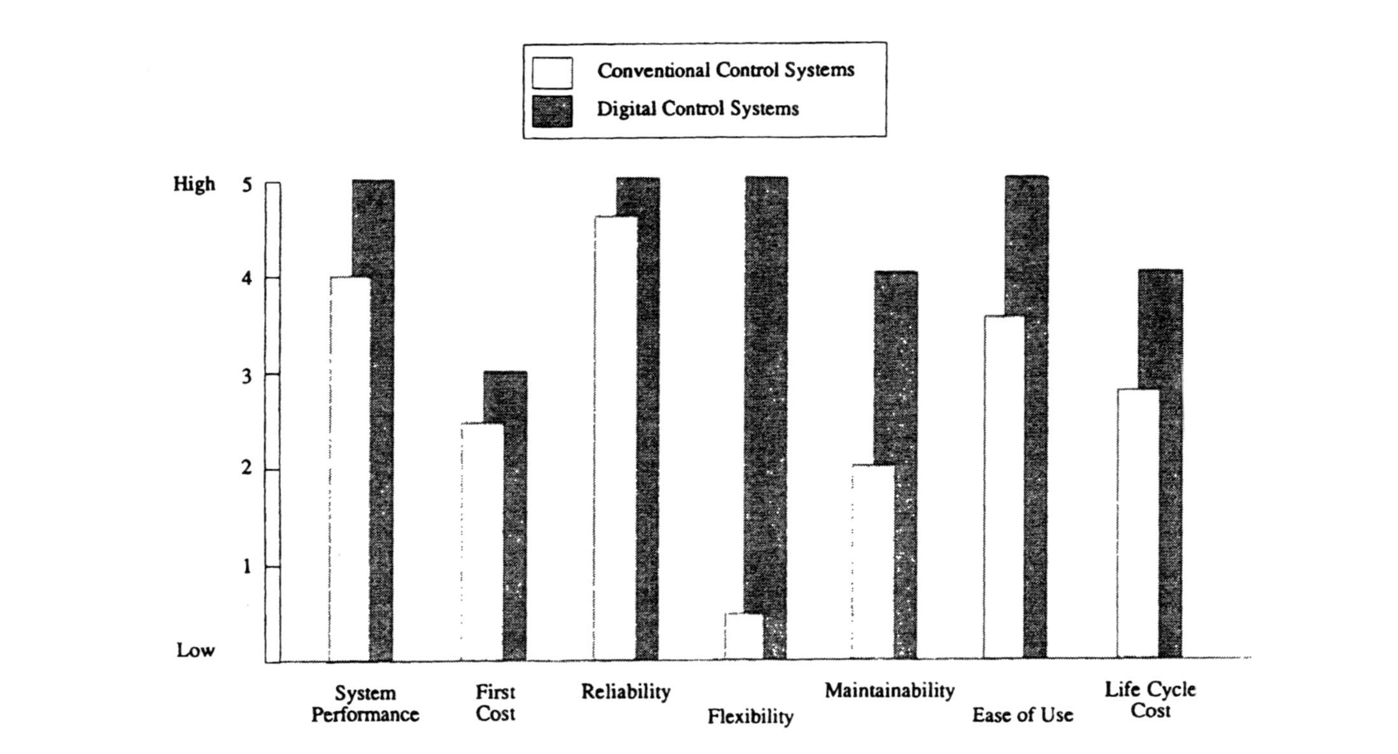

Figure 1-3 Conventional and digital control system comparison.

can add value to a property and provide property managers with an excellent tool for controlling building operating costs and for apportioning tenant expenses after a building is occupied.

The domestic market for commercial temperature control systems exceeded US$ 8 billion in 1996. In the 1970s, a small group of large companies manufactured temperature control equipment. Today, more than 100 companies produce intelligent energy management and control products for the commercial building market, ranging from intelligent time clocks to fully integrated environmental, life safety, and security systems capable of controlling multisite office complexes.

As with any industry that is going through a period of rapid innovation, there is change taking place in the competitive balance of the controls industry. Whereas in the past such industry giants as Honeywell, Johnson, and Powers dominated the manufacture of control devices, microprocessor technologies have made it possible for many smaller firms to compete for market share. In fact, some smaller firms have done more to innovate and apply new technology than their larger brethren. So effective have been their innovations that many have been acquired by the industry majors to bolster their own product development efforts.

Another form of innovation has been joint marketing efforts between major manufacturers of HVAC equipment and control systems. Companies such as Trane and Carrier, which primarily produce HVAC equipment, have entered into private marketing agreements with major controls manufacturers to "bundle" factory-installed DDC controls with their HVAC equipment. The concept is that factory labor is less costly than field labor, and one source of system responsibility is attractive to contractors and building owners. As competition in the building systems industry continues to grow between the major suppliers, more joint marketing efforts of this type will take place. However, it remains to be seen whether such alliances are truly effective in lowering the total cost and improving the performance of HVAC systems.

Future of Direct Digital Controls

The field of automatic control is undergoing the most dramatic change it has seen in the last 50 years, when electric control systems were first introduced to the commercial HVAC market. Modern control systems combining elements of electronic, microcomputer, and telecommunication technologies are being introduced to meet the increasing demands of the commercial building industry. Conventional control systems utilizing fixed-system architectures are facing obsolescence as fast-paced innovations in microprocessor-based control products are challenging the cost effectiveness of traditional pneumatic and electric controls. Moreover, even sophisticated digital systems using proprietary programming and communication technologies face obsolescence as the industry evolves toward "open architectures" that offer building owners independence from a single source of supply.

Smaller, modular, and more intelligent control devices are being manufactured for the HVAC industry than ever before. The capability to control a single device, such as a variable-air-volume (VAV) terminal or a small fan coil unit, is now possible at the local level. Communication between intelligent independent controllers in a building offers design engineers a great deal of flexibility in both new construction and in retrofit applications. In the future we will continue to see computer intelligence drilled down to increasingly lower component levels as cost and physical size limitations are overcome.

The next step in the evolution of intelligent digital building controls is the development of a protocol that will allow control systems of dissimilar manufacture to communicate and share operating information in a real-time format. Once the topic of much speculation and debate, this "handshake" among proprietary systems is quickly becoming a reality as industry groups and controls manufacturers work together to develop standards. Although in the recent past manufacturers paid only lip service to the benefits to building owners that such a protocol could offer, today they are embracing these reforms in response to overwhelming market demand for interoperable or "vendor-independent" systems. "Conform or be cast out" is the message the market is sending, and digital system manufacturers are listening. This issue, along with the most promising technologies emerging in this area, are discussed in Chapter 5.

The future will also bring new levels of intelligence and user operability to building HVAC controls. The technology to make control systems intuitive is available, proven, and affordable. More than two decades have passed since the first direct digital controllers were produced for the commercial building market. Like any technology associated with the computer industry, there have been quantum changes in the size, shape, and speed of these machines over a relatively short time.

As large-scale integration of computer chips continued to develop, the CPU boards that were the engines for these systems became smaller, more compact, and more powerful. Rapidly developed generations of microprocessor-based controllers continually became smaller, less expensive, and easier to program as software developers and equipment manufacturers embed this technology in their products. Today, microprocessor intelligence resides at every level of a commercial control system, from the room sensor that measures changes in temperature or motion sensors that detect occupancy all the way up to the computers that monitor the behavior of many systems within a campus of buildings. The intelligence is everywhere. The trick is to connect these units of remote intelligence into a system that saves energy, makes buildings easier to manage, and keeps occupants comfortable.

By 1997, there were more than 200 million personal computers in operation in the world. However, there were more than 6 *billion* microchips embedded in mechanical and electrical control devices ranging from microwave ovens, VCRs, and camcorders to surgical instruments and commercial HVAC control systems. Intel Corporation estimates that the number of personal computers in use worldwide will grow to more than 500 million by the year 2002. However, the number of embedded microchips is expected to grow at a rate far beyond this as manufacturers find new ways to differentiate their products using digital intelligence.

With all of this intelligence so widely distributed at every level of a building HVAC control system, our challenge will be to not only optimize its local operating performance but also to harness this collective intelligence to create systems that render better living and working environments. The technology is here. It's up to use to make the most of it.

The chapters that follow begin with an overview of basic control system operations and how microprocessors work, and then move on to how they are connected into working systems, how they communicate to their human hosts, and finally how to apply them to real HVAC systems.

Chapter 2

Fundamentals of Control Systems

The purpose of automatic controls is to modulate the capacity of HVAC equipment to satisfy the heating and cooling requirements of a building and to monitor and control safely the operation of HVAC systems. This type of control traditionally has been performed by pneumatic and electric control systems. Today, microprocessors are taking the place of conventional controllers and are rapidly becoming the preferred automatic control method for building HVAC systems. Because of the declining cost of semiconductor chips and the proven reliability of microprocessor-based controllers, direct digital control (DDC) systems now compare favorably against conventional control methods in both functional and economic terms. DDC, however, is not the answer to every automatic control problem. A solid grounding in control system fundamentals is necessary before one can fully understand how to best apply direct digital controllers to common control system designs. Because DDC systems are typically used in combination with other types of control systems, control system fundamentals are now reviewed.

Terminology

Control theory is the field of study concerned with the way that components in a system interact to perform as a functional unit. A control

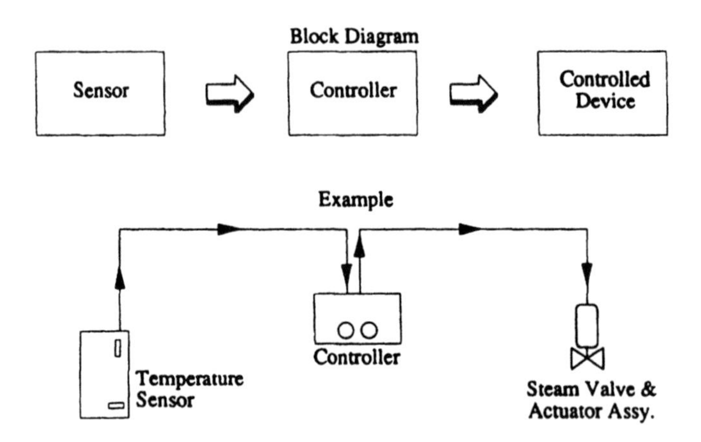

Figure 2-1 Basic control system.

system comprises three basic elements: a sensor, a controller, and a controlled device. A block diagram of a basic control system is shown in Figure 2-1. Sensors measure a controlled variable, which can be temperature, pressure, or humidity. Current information on the state of the controlled variable is transmitted to a controller, which interprets these data against a predetermined value called a set point, which is the desired state of the controlled variable. When the controller detects a change in the controlled variable, it sends a corrective response to a controlled device, which will act to return the controlled variable to its set point. Figure 2-2 depicts a simple control system operating a steam radiator. A sensor in a room measures the tempera-

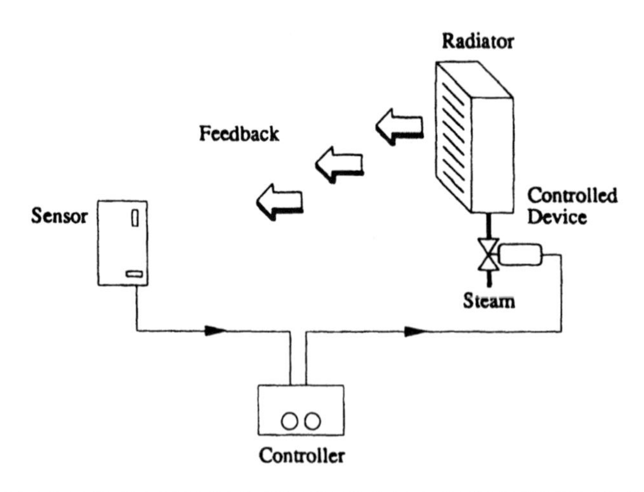

Figure 2-2 Closed loop control.

ture of the space. The desired temperature for the room, or set point, is 70°F. When the sensor informs the controller that the room temperature has fallen below the set point, the controller opens the steam valve until the room temperature returns to 70°F.

Feedback is corrective information about a condition or an action that is returned to a point of origin. There are two basic control relationships, known as closed-loop systems and open-loop systems, that are distinguished by the presence or absence of feedback.

Closed-loop control systems have the ability to self-correct for changes in a measured variable by means of feedback that travels along a closed path. This feedback is continuous and therefore allows precise control of the controlled variable. Figure 2-2 is a good example of a closed-loop control system. As the room temperature falls, heat is added to the space when a steam valve is opened, thus allowing the radiator to heat the space; as the space temperature returns to the set point, the controller closes the steam valve. This self-correcting action is the essence of automatic control.

Open-loop control systems do not provide this self-correcting action because they act on indirect rather than direct information. For example, the radiator in Figure 2-2 could be turned on by a time clock, instead of a thermostat, from 6 a.m. until 8 a.m., because we assume that the room will be cold. The room, however, may be warm, but the control system has no way of reading the space temperature. Obviously, closed-loop automatic control systems are preferable for HVAC systems requiring uninterrupted control of a measured variable, and therefore they are the mode of control most commonly used in HVAC applications.

Communication between the elements of a control system is accomplished in two basic forms: digital and analog. Digital, or binary, information is expressed in two states, positive or negative, indicating that a measured variable has reached a predetermined limit, for example, on/off, open/closed, flow/no-flow. Therefore, a digital control signal indicates either a 100% or 0% condition. The terms *digital* and *binary* are taken from the binary, or base 2, system of numbers, which is comprised of the digits 0 and 1. Typically, the off state equals 0, and the on state equals 1.

Analog signals represent information in measurable increments that are continuously changing or modulating. For instance, a thermometer measuring temperature by gradually rising and falling in proportion to the amount of heat it senses is considered to be analog. The hands on a clock are another familiar example of analog indication.

Control loops are organized groups of control system elements, or blocks, that are drawn in diagrammatic form for the purpose of modeling and evaluation. Control loops are usually presented in block format for simplicity and ease of understanding. Figure 2-1 shows a block diagram of a single-control loop, showing the flow of information between components in the loop. A single loop consists of one block of incoming information, one block of decision making, and one block of corrective response. Groups of control loops can be linked to create control sequences to meet virtually any control system requirement, and they are the foundation of control system design. The terms *loop* and *block* are used many times throughout this text, especially in discussions of computer-based control programming and the creation of custom control algorithms. Transfer of information between the blocks can be expressed mathematically by first-order differential equations; however, proper treatment of this subject is too encompassing to include here.

Control System Performance

Automatic Control Actions

Automatic control is the ability of a system to self-correct output signals to a controlled device in response to constantly changing input signals from a controlled variable. Automatic control actions are generally categorized into five modes, which are differentiated by their response rate to changes in the controlled variable and the accuracy with which they return the controlled variable to its desired condition. The modes are:

- Two-position
- Proportional
- Proportional plus integral (PI or reset control)
- Proportional plus integral plus derivative (PID)
- Artificial intelligence (AI)

Two-position control, also known as on/off control or digital control, is the simplest of the four control modes. A system controlled in this manner is always floating between the off state and the on state. The performance of this mode depends on the distance between the on/off states and how fast the system will respond to changes in the measured variable. Imagine taking a shower by first turning on the hot water 100%, then the cold water 100% in an effort to achieve a comfortable water temperature. The level of comfort you achieve depends on how fast you can alternate between hot and cold.

Figure 2-3 illustrates the response of a two-position control action. Notice the oscillations between the on and off limits and the effect on the controlled variable over time. The rapid cycling between on and

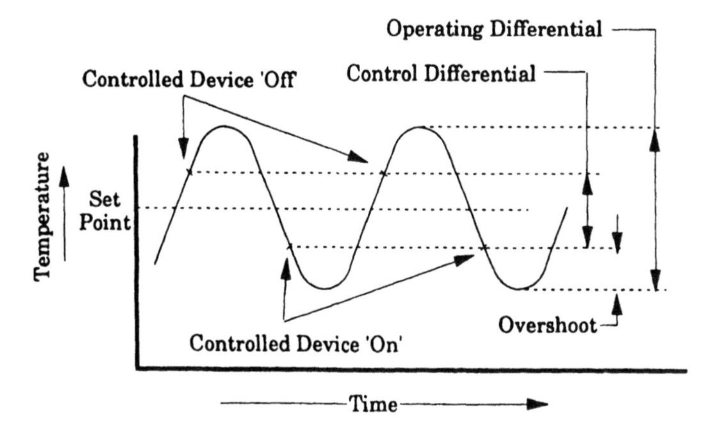

Figure 2-3 Two-position control action chart (heating action shown).

off states creates a problem known as hunting, whereby the system struggles to gain control by continually moving between on and off states, or short cycling. Because repetitive starting and stopping increases wear and tear on HVAC equipment, a limit is introduced to slow this cycling and bring the system into control. This limit, also known as differential, is simply the difference between the on and off states expressed in terms of the measured variable, such as degrees of temperature. Adjustment of the differential can improve the performance of a control system by reducing the overshoot or swing that is caused by cycling.

The operating differential of the system, which is the differential expressed as the actual effect on the measured variable, is somewhat higher than the on/off differential because of this overshoot.

A method of reducing such excessive swings in the controlled variable is called anticipatory two-position control. By introducing an artificial differential on the system, we can anticipate the amplitude of the swing and offset it. The result is a much-reduced operating differential and a system that is in better control (see Figure 2-4).

Floating control is a form of two-position control that requires a special controlled device that can start and then stop in midstroke while maintaining its position between the on and off states. Figure 2-5 charts the typical action and response of a floating-control loop. Suppose that a chilled-water valve with a floating actuator provides cold water to a coil that cools air entering a room through a duct. A room thermostat signals the valve actuator to open or close based on changes in room temperature, allowing more or less cooling to occur. When the desired room temperature is achieved, the thermostat stops the valve in midstroke and holds it in its present position until it is

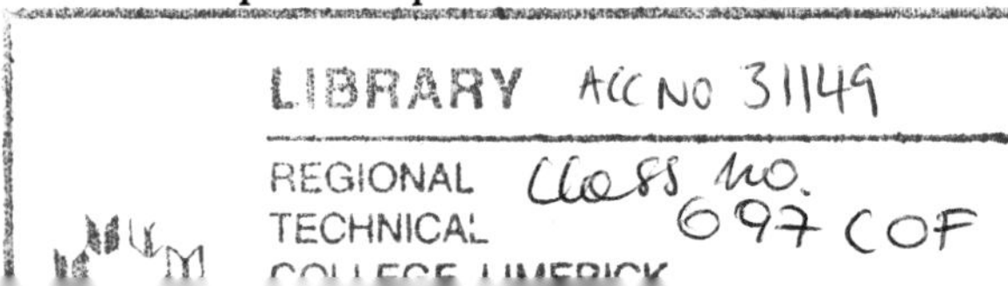

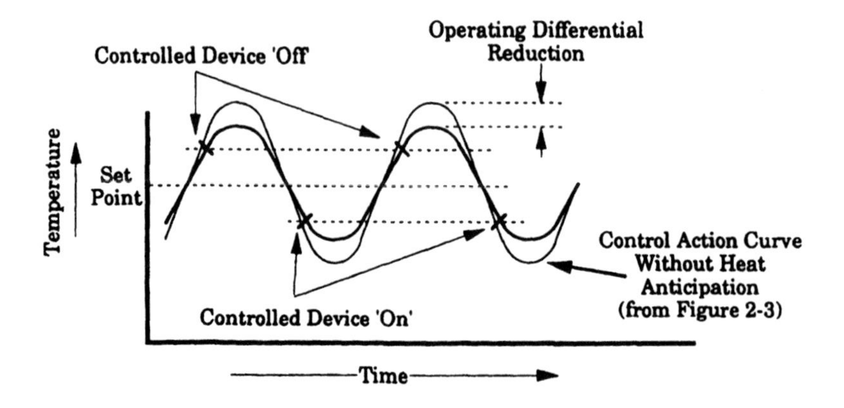

Figure 2-4 Anticipatory two-position control action chart.

again used to provide chilled water. The valve may never fully open or close; it simply "floats" between the two extremes. Floating-control devices are usually designed to have a long stroke, or period of time, between fully open and fully closed positions. This allows for changes in the controlled variable to occur before full stroke is achieved, thus reducing the overshoot of the system.

Proportional control means that the output of the system can vary incrementally over the entire range of its control. Unlike the abrupt start/stop of floating control, a controlled device can be proportionately positioned in response to slight changes in the controlled variable, thereby satisfying the control objective in the shortest time and with the most accuracy.

There are several important terms associated with proportional control. Refer to Figure 2-6 for a graphic representation of this control action. *Throttling range* is the total amount of change that occurs in

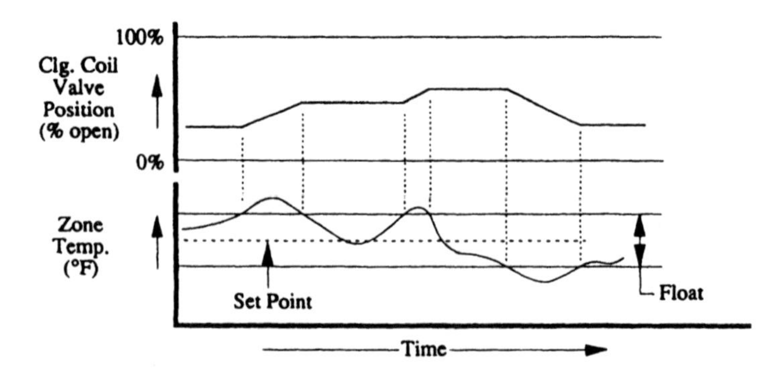

Figure 2-5 Floating-control action.

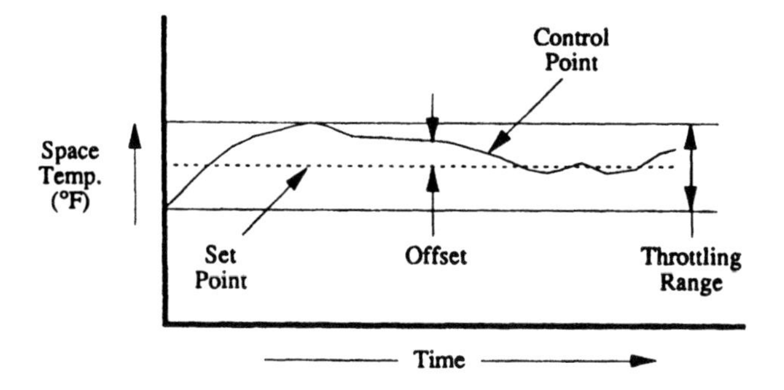

Figure 2-6 Proportional control action.

the controlled variable for the controller to modulate the controlled device over its entire range of control. *Set point* is the desired value of the controlled variable. The controller is programmed to maintain the controlled variable as close to the set point as possible. In reality, the actual value of the controlled variable, or control point, is slightly different than the set point because of inaccuracies in the performance of the control system. This difference is referred to as offset, also known as the deviation or droop of the system.

Figure 2-6 illustrates variations in the controlled variable over time. When you compare two-position and floating-control strategies against proportional control, it is obvious that proportional control is much more accurate. The maximum offset value in a proportional control action will be less than half of the operating differential of two-position control modes.

Proportional plus integral, also known as reset control or PI control, is identical to the proportional control mode, except that a new function is introduced (see Figure 2-7). The integral function is used to eliminate offset by changing the control point of the controller by the amount of

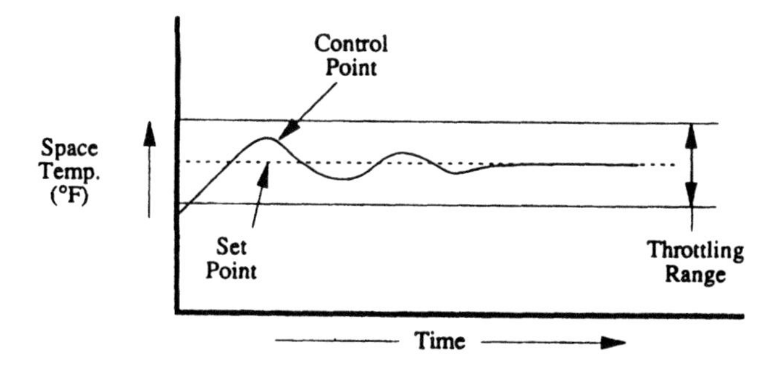

Figure 2-7 Proportional plus integral (PI) control action.

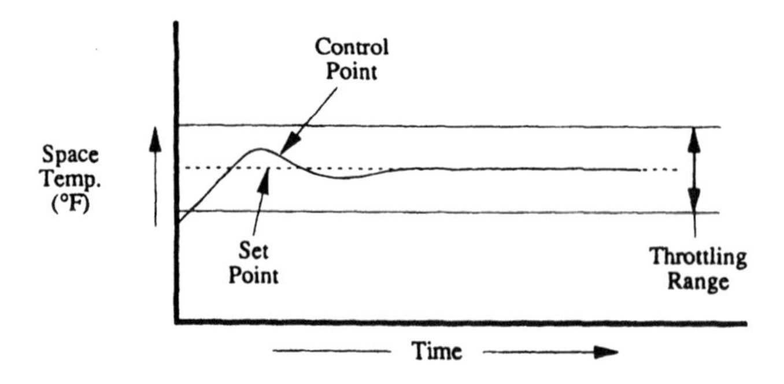

Figure 2-8 Proportional plus integral plus derivative (PID) control action.

the offset to bring the control point back to the set point. The number of times that this reset action takes place in a given time frame is called the reset rate. Proportional plus integral, or PI, control is commonly used in systems in which changes in the loads affecting the controlled variable are slow acting, which is typical of building HVAC systems. Proportional plus integral plus derivative (PID) control adds yet another element to the proportional control equation: time. The derivative function controls the speed with which the integral function resets the control point. Figure 2-8 overlays the derivative control action on the PI control mode.

Because it is microprocessor-based, a digital controller is capable of fine-tuning an HVAC system by using PID control functions. For example, automatic reset control, which is available in conventional devices, can maintain the set point of a stable proportional control system. However, the time necessary to restore the set point after a large upset in the building load may be substantial or objectionable in certain applications. DDC devices with PID control will greatly reduce this time delay. PID control can improve the response of a control system by decreasing the time lag between an environmental change and the control system response.

Three values are necessary to achieve a well-tuned PID control system. Range, reset, and rate directly correspond to the proportional, integral, and derivative functions. The proportional segment of PID responds like any conventional controller to maintain a stable condition within the controlled space. The proportional control function operates within the throttling range discussed earlier.

Range corresponds to the system throttling range or proportional band on a conventional controller. A range that is too narrow will cause instability in the execution of control routines. A wider range will provide for the stability necessary for the next two adjustments. The integral segment of PID controls system response by sliding the throttling range window up or down a value scale. The digital controller determines the amount of change necessary to alter the position of the control device enough to remove any offset or droop. Without integral control, the time needed to restore set point after a load upset is usually close to the same amount of time required for the proportional control to establish a stable condition. In some cases, this time delay, or transport lag, is so long that the offset is not removed before another load upset occurs.

Reset is the adjustment of the integral action timing. One repeat is the time necessary to duplicate or repeat the proportional action that occurred as a result of an upset of space temperature. In general, reset time should be adjusted with the system in operation and under some load. A reset time that is too short will result in instability around the set point, eventually forcing the system out of control. A reset time can be too wide, but a wider time at least ensures stability. It is not possible to use integral action without proportional action.

The derivative section of PID senses the rate of upset. In derivative control, the microprocessor controller drives the control device above or below the integral action long enough to bring the rate of change of the controlled variables to zero. As set point is approached, this override of the integral action tends to reduce the time for the system to restore set point. It also reduces the time delays caused by transport lag or thermal integral. Another advantage is that the derivative action greatly reduces and often eliminates overshoot of set point because the rate of change is reduced to zero as the system approaches the set point value. If another upset occurs quickly, the chances are greater that the next upset will not happen during the time when set point is being restored. Upsets that occur during the stabilizing process lead to longer and larger instabilities in a control system.

Derivative action is modified by adjusting the rate. The rate value is the amount of change per minute in the controlled variable necessary to produce a 100% change in the output of the controller. For this reason, the rate should be no smaller than the optimum system throttling range or proportional band.

Although PID control is certainly an improvement over proportional control, it is not ideally suited for most HVAC control applications. PID control theory is based on the assumption that a linear relationship exists between the inputs and outputs of a controller. Nonlinear HVAC control applications, such as the regulation of water flow through a valve to a coil, are difficult to control with PID because it cannot compensate for the nonlinear characteristics of a controlled device without compromising control accuracy. For instance, the performance curve of a control valve is nonlinear, as shown in Figure 2-9. The valve is usually downsized from the line size to increase the

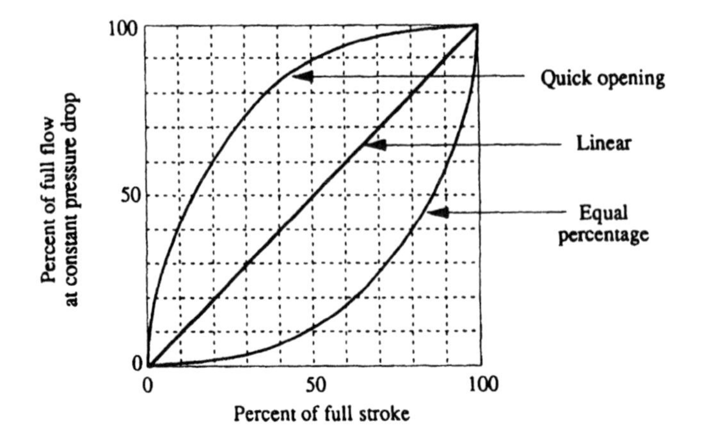

Figure 2-9 Control valve flow characteristics.

pressure drop across it. This is done to provide the controller with a wide proportional band with which to control the modulation of the valve. The PID controller assumes that its input and output relationships are linear; therefore, this differing control characteristic must be offset by using pressure drop. If the controller output signal could modulate with respect to the flow curve of the valve, the valve could then be line-sized, thus reducing the pressure drop across it and ultimately reducing the overall pumping pressure required to operate the hydronic system.

There have been significant developments in the area of nonlinear or "expert" systems that are capable of "learning" the characteristics of the systems they control. These developments are discussed in Chapter 11.

An artificially intelligent (AI) control loop can be "taught" the specific characteristics of each device under its control and can control simultaneously many devices that have dissimilar characteristics. This method is far less restrictive than conventional control methods, which can only control devices that have similar characteristics. The ability to tune each control loop depends on the capacity of the controller; this is where the microcomputer makes its greatest contribution to the HVAC control industry. The application of microcomputers in HVAC control systems is explained in the next chapter.

AI is a self-learning form of control, whereby a controller "learns" how to anticipate changes in controlled variables by evaluating past and current control system performance. This intelligence resides in software programs operating within the DDC system microprocessor. By using computer language that can manipulate both symbols and numbers, AI software can mimic human thought behavior to solve

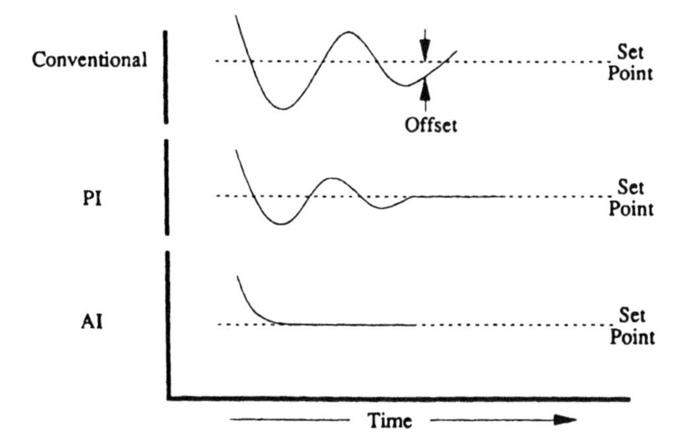

Figure 2-10 Response rate and accuracy comparison.

control problems by observing HVAC system behavior, storing this information in its database of knowledge, and using these data to make anticipatory control decisions.

Logical assumptions drawn from its database about the behavior of the building load allow the AI system to act in a predictive and preventive fashion to anticipate problems before they actually occur. AI is currently being developed for HVAC control applications to address the need for more accurate forms of loop control. Figure 2-10 compares the response rate and accuracy of conventional, PI, and AI control methods.

Control System Energy Sources

There are four primary sources of energy for control system components:

- Pneumatic
- Fluidic
- Electric
- Self-powered

Pneumatic control systems are powered by compressed air at 15 to 20 psig, although some special pneumatic control devices, such as large damper actuators, may require 100 psig air to operate. Mechanical controllers modulate the pressure of the compressed air in response to changes in the controlled variable to produce corrective output responses to controlled devices. Pneumatic control systems have been popular for many years because they are easy for building operators to understand and traditionally have offered the lowest first cost.

Fluidic systems are similar to pneumatic systems, but they use a gas or fluid other than air as the control medium. The operation of a hydraulic mechanism is an example of a fluidic control system.

Electric energy is used for both electric and electronic control systems. Electric controls use line voltage, 120 volts alternating current (VAC), low voltage, or 24 VAC, and are two-position, with the electric controller passing or interrupting voltage to start or stop an electric motor. Electronic controls utilize low-voltage power, usually 24 VAC or less, and use electronic circuits to sense and transmit varying voltages for the proportional control of controlled devices.

Self-powered systems combine a sensor, a controller, and a controlled device into one unit and do not require an outside source of power to operate. Changes in the controlled variable result in a change in the pressure or volume of an enclosed medium, which provides energy to a controller to produce a corrective output response. A familiar self-powered system is the self-contained steam or hot-water valve found on old steam radiators.

Control System Elements

As stated earlier, all control systems, no matter how complex, comprise three elements—a sensor, a controller, and a controlled device. Any device found in a control system will fall into one or more of these three major categories.

Sensors

A *sensor* is the most critical element of a control system because it measures the current value of a controlled variable. Controlled variables include temperature, humidity, pressure, enthalpy and conductivity. A change in the controlled variable produces a change in the physical or electrical properties of the sensor that can be translated into useful information.

To achieve accurate closed-loop control, changes in a controlled variable must be (1) measured, (2) transmitted to a controller for interpretation, and (3) acted on to produce a corrective response. There are four categories of sensing devices used in automatic control systems that are differentiated by the type of control action they perform and by the degree to which they integrate the functions of sensing, transmitting, and controlling:

- Sensors
- Transmitters and transducers
- Sensor–transmitters
- Sensor–controllers

Sensors measure the current state of a controlled variable in either an analog or digital method. Analog, or proportioning, sensors recognize changes in the measured variable in small increments and continuously transmit varying signals in proportion to the changes in the measured variable. For instance, a resistance sensor may vary its signal 2 ohms for every 1° change in temperature. Digital sensors indicate a change in the controlled variable by means of a contact closure. A switch to indicate water flow in a pipe will move from the off position to the on position when a pump starts. A contact will close, allowing current to pass through a circuit to indicate to the controller that the pump is on. Digital states can be monitored for a variety of operating conditions, such as equipment operation status and position indication. An airflow switch, for instance, tells the controller whether a fan is operating. A position indicator, such as an actuator end switch, tells the controller that a controlled device has reached the maximum or minimum limit of its stroke.

Transmitters are special-purpose devices that amplify low-intensity signals from sensors and transmit these signals over wires or tubes to a controller. The distance that a control signal can travel over wires or tubes is limited because of resistance losses (wire) or friction losses (tube) caused by the medium. Transmitters ensure that a clear, strong, accurate control signal is presented to the controller. The distance between a sensor and a transmitter is kept as short as possible, but the distance between the transmitter and the controller can be quite long.

Transmitters are available in pneumatic and electronic form and are commonly used in temperature, humidity, and pressure control applications.

Transducers translate a control signal from one form to another, such as pneumatic to electronic, or analog to digital. They are commonly used in systems designed to combine the benefits of different types of control systems or where dissimilar control signals must be integrated into a single control system.

Sensor–transmitters are the combination of a sensor and a transmitter in one package.

Sensor–controllers are the combination of a sensor and a controller in one package. A thermostat is a popular example of a sensor–controller. A thermostat controls the environment in a space by sensing the temperature and then modulating a device, such as an air damper, to maintain the room temperature set point.

There are five specific types of sensors. They are grouped according to the environmental variables they are designed to measure:

- Temperature sensors
- Humidity sensors
- Pressure sensors
- Flow sensors
- Special-purpose sensors

Temperature Sensors

The most commonly measured variable is temperature. There are seven types of temperature-sensing elements, classified according to the method by which a change in temperature is detected and communicated to a controller:

- Bimetal
- Rod and tube
- Sealed bellows
- Remote bulb
- Thermistor
- Resistance temperature detector (RTD)
- Thermocouple

Bimetal sensing elements are created by joining two strips of dissimilar metals that have different coefficients of expansion. A known rate of deflection occurs between these two metals during a change in temperature. If the sensor is calibrated to a datum temperature of, say, 70°F, a known rate of deflection can be interpreted to indicate a linear change in temperature. Bimetal elements are used in electric and pneumatic sensors in both two-position and proportional control actions. Pneumatic transmitters use bimetals to modulate the release of air pressure to a controller to indicate a change in temperature, causing a proportional change in the output pressure to a controlled device. Pneumatic transmitters are widely used, especially in conventional control systems and for the control of variable air-volume terminals for zone temperature control as shown later. They are easy to understand and are relatively inexpensive; however, their accuracy will drift over time, and they require periodic maintenance to operate properly.

Rod-and-tube elements are frequently used for immersion- or insertion-type sensors. A rod made of low-expansion metal is inserted into a tube composed of high-expansion metal; the inner tube changes length in proportion to changes in temperature, which allows the rod inside it to move. The movement of the rod indicates a change in the temperature of the controlled variable.

Sealed-bellows elements are filled with a vapor, gas, or liquid devoid of air. A change in temperature results in a change in the pressure or volume of the medium enclosed in the bellows. Sealed-bellows elements are sometimes found in older models of room thermostats.

A *remote-bulb* sensor is similar in design to a sealed-bellows sensor. A bulb is attached by a capillary tube to a bellows, and the entire assembly is filled with a vapor, gas, or liquid devoid of air. Changes in temperature cause a change in the volume of the enclosed medium, which travels from the bulb through the capillary to the bellows. Remote-bulb sensors are frequently used in applications in which the controlled variable is located remotely from the controller. Rod-and-tube, sealed-bellows, and remote-bulb elements are common to pneumatic and fluidic control systems and are generally accurate to within 2% over their range of measurement.

A *thermistor* measures temperature in the form of electrical resistance. As a measured temperature changes, a representative change occurs in the electrical resistance of a semiconductor material. The resistance curve of a thermistor is nonlinear over its range, which means that the ratio of change between electrical resistance and temperature is not 1:1. For example, a 1° increase in temperature may produce a difference of 3 ohms of resistance. This ratio is known as the coefficient of change.

Thermistors are usually available in resistances of 10,000 ohms or higher. Such a wide range is beneficial in critical control applications because it provides a large change in resistance for a small change in temperature. Much finer control can be accomplished when the coefficient of change is a large multiple of the change in the controlled variable. Because of their precision and relative low cost, thermistors are often used with computerized control systems.

A *resistance temperature detector* (RTD), like the thermistor, uses the electrical resistance of a metallic material to indicate a change in temperature. Materials used in RTDs are selected for their linear resistance characteristics when exposed to a variable temperature range, and include platinum, copper, tungsten, nickel, and iron alloys. The difference between these materials lies in their degree of linearity over a given temperature range. Platinum, for instance, is considered to be a highly accurate RTD material because it is linear from 0 to 300°F with a tolerance of 0.3%. Unlike the thermistor, RTDs are usually combined with an integrated circuit to produce a current signal over a finite control range. The 4- to 20-milliamp (mA) current output RTD is a familiar example. RTD sensors are available in different housings for applications in occupied spaces, surfaces, ducts, immersions, and outside air, and are generally favored for their low cost and high degree of accuracy.

A *thermocouple* sensor consists of two dissimilar metal wires joined at two points. These two points, called junctions, create the thermocouple circuit. While one junction is kept at a constant temperature the other is allowed to vary in relation to temperature changes in the controlled variable.

An electric current flows through the thermocouple circuit relative to the voltage potential of the two junctions; the voltage of the thermocouple circuit is then measured to indicate temperature. A 4- to 20-mA direct current signal is the output from thermocouple transmitters. Because thermocouple circuits are not truly linear owing to the characteristics of the metal wire, modern thermocouple devices have been combined with linearization circuitry to linearize the output signal over the entire range of the sensor. Thermocouples are used primarily for high-temperature control applications and are seldom applied to HVAC systems.

Humidity Sensors

Humidity sensors measure the dew point, or relative humidity, of an airstream. Humidity sensors use hygroscopic materials that respond physically or electrically to changes in the amount of atmospheric moisture present.

Hygroscopic elements are mechanical in nature, in that a hygroscopic material exposed to moisture will expand. The expansion of the sensor material is detected by a mechanical linkage and then converted into a useful control signal.

Electronic humidity sensors were designed to be used with computerized control systems, energy management control systems, and process control systems. They detect measurable changes in the resistance or capacitance of selected materials under changing moisture conditions, which are converted to indicate relative humidity. Because water is an efficient electrical conductor, a grid coated with a hygroscopic material exhibits a conductance in direct relation to the amount of moisture present. Most electronic humidity elements are composed of highly sensitive metal oxide ceramic disks that are electrically charged. The rate of resistance to the flow of electricity between the disks produces a signal that can be converted into a voltage or current output, such as the familiar 1 to 5 VDC or 4- to 20-mA output signals used with computerized control systems. Electronic sensors are highly accurate, but they are not linear and therefore require modification to the output signal to linearize their response. Modern humidity sensors are designed with on-board circuitry to correct the output response for changes in ambient temperature and to linearize the signal.

Pressure Sensors

Pressure sensors convert changes in absolute, gauge, and differential pressures into a mechanical motion that can be measured by a diaphragm, bellows, or Bourdon tube. Pneumatic pressure sensors produce a change in air pressure in response to a change in the measured pressure.

DDC systems frequently use electronic pressure sensors for their low cost and high degree of accuracy and signal repeatability. Electronic pressure sensors produce a variable voltage or current by converting the mechanical motion of the pressure-sensing device into an electrical resistance. An electronic circuit corrects the sensor signal for changes in ambient temperature. Because measured pressures can be very low, this circuitry also amplifies the signal so that it can be easily read by the controller reading the signal. Pressure sensors can be of the two-position type, such as a differential pressure switch, which indicates an excessive pressure drop across a dirty air filter, or of the proportioning type, which monitors pressure conditions on an ongoing basis.

Flow Sensors

There are several methods used to sense the flow of a gas or liquid, and each is different in terms of its range of control, accuracy, materials of construction, and, ultimately, cost. Pitot tubes, venturis, turbines, magnetic meters, and vortex meters are common devices used to measure flow in HVAC control systems. However, the most commonly measured medium in HVAC systems is airflow.

A popular energy conservation strategy used in HVAC systems is variable air volume (VAV). VAV control is based on modulating the speed of supply and return fans in sequence to provide only enough air to meet the load of the occupied space. To achieve this control, the volume of air passing through the system must be measured accurately. An airflow measuring station is a device that measures the velocity pressure of a system and converts this value into volume, which is expressed in cubic feet per minute. The design of airflow measuring stations is based on a self-averaging pitot tube array that traverses the cross-sectional area of the duct to collect multiple pressure signals which are then averaged to yield a velocity pressure signal. This is done by measuring total pressure and static pressure and then subtracting one from the other to derive the velocity pressure. Air velocity, in feet per minute (fpm), is a function of the velocity pressure, which is converted by the formula

$$\text{Air velocity (fpm)} = 4005 \text{ velocity pressure}$$

The velocity pressure signal from the airflow measuring station is passed through a square-root extractor and then multiplied by the constant 4005 to yield the air velocity. Air velocity multiplied by the area of the duct will render the air volume of the system in cubic feet per minute. Modern airflow measuring stations incorporate the pitot tube arrays, square-root extractor, controller, signal conditioner, and output signal transmitter into a single assembly. A critical component in the airflow measuring station is the square-root extractor, which converts velocity pressure signals into air velocity signals. A discussion of square-root extractors is given later.

Special-Purpose Sensors

A diverse group of sensors used to indicate environmental conditions and measure energy consumption are available.

Power Measurement Devices

Electrical consumption sensors are used to measure the amount of electric energy consumed by a system. They are commonly used in DDC systems to measure the overall effectiveness of an energy management program. Power measurement is also used to apportion system operating costs to building tenants in accordance with their individual rates of power usage. There are three common devices used to measure power consumption: watt transducers, current transducers, and kilowatt pulse meters.

Watt transducers measure the current and voltage of a circuit and provide an output signal that is proportional to the amount of power passing through the circuit.

Current transducers measure the current in a circuit and provide a current output signal that is proportional to the measured current.

Kilowatt pulse meters emit an electrical pulse that is proportional to the power passing through a circuit. The digital controller counts the quantity and frequency of pulse to determine the rate at which a device is consuming kilowatt-hours.

Life Safety Devices

Smoke detectors are the primary sensing element in fire a[illegible] fe safety control systems. They are two-position devices that utilize [illegible]ff relay contacts to indicate the presence of combustion products. Although these sensors are part of the smoke management and life safety systems, they are commonly used in HVAC systems to initiate smoke control operating sequences. In the early stages of a building fire, the

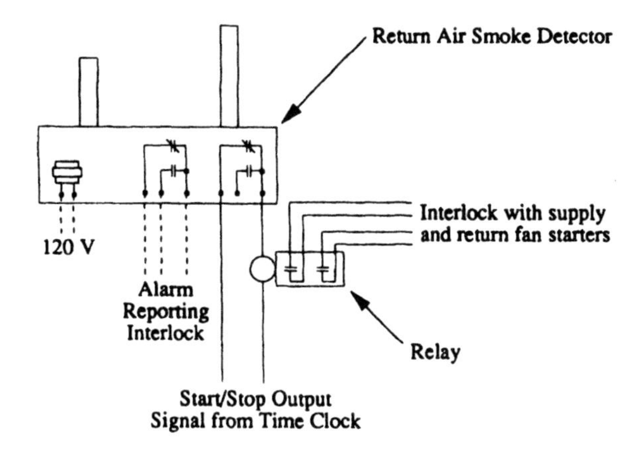

Figure 2-11 Smoke detector interlock wiring.

HVAC fan system can be used to pressurize and contain smoke while the controller annunciates an alarm. Figure 2-11 illustrates the standard method of wiring a smoke detector to interface with both an HVAC control system and a building fire and life safety system.

The objective of a smoke detection system is to detect an impending fire before it reaches the incendiary stage. Because the greatest threat to life in a fire is from asphyxiation, combustible gases must be quickly detected before they reach a lethal quantity. There are two primary types of smoke detectors used for this purpose: ionization and photoelectric.

Ionization smoke detectors employ a small ionization chamber that senses the invisible combustion gases present in the early stages of a fire. These combustion gases affect the electrical conductivity of the chamber by lowering the amount of current that can pass through it, thereby indicating the presence of products of combustion. An electronic circuit is then closed to initiate an alarm sequence. Ionization detectors respond well to small particles that are generated by hot, smoky fires. The smoke detectors found in homes are of the ionization type, mainly because of cost and the fact that most home fires will occur close to the smoke detector.

Photoelectric smoke detectors employ a light beam and a photoelectric cell. As combustible gases pass through the light beam, they reduce its intensity until a critical value is reached, causing an alarm sequence to occur.

Photoelectric smoke detectors are most applicable in life safety applications because they are more responsive to smoldering or low-grade fires, which produce dense smoke. As smoke moves away from its source, it agglomerates into larger particles. This agglomeration is caused by the cooling effect of the air on the smoke. Generally, a photoelectric smoke detector will respond faster to large smoke particles than will ionization-type detectors.

There is no standard guideline that dictates which type of detector to use in every application. Therefore, the decision to choose either the ionization or photoelectric method must be carefully weighed against the kind of environment the detector must protect, as well as local building and fire code requirements. It is a good practice to check with local building officials and the local fire marshall for guidance on which method of protection is preferred in your area.

Other Special-Purpose Sensors

Carbon monoxide (CO) detectors are used in enclosed parking facilities to indicate excessive levels of carbon monoxide emissions from automobiles. They are frequently used to operate garage exhaust fans to evacuate these harmful emissions.

Occupancy, or *motion, detectors* are used to curtail energy consumption by controlling HVAC services to building areas that are infrequently occupied. In libraries and conference rooms, for instance, HVAC services may be started whenever these areas become occupied and shut off when they are evacuated. They are also used in security systems to warn of an unwanted intrusion.

Sensor Selection Considerations

The proper selection of a sensor depends on the specific control application requirements, the power source available to operate the control system, and the degree of accuracy required.

A control system can operate only as accurately as the sensors used to measure the controlled variable. Differences in occupancy and functions within buildings demand different degrees of control system performance and accuracy. For example, an acceptable temperature drift in a commercial office environment might be 3 to 4°F; such drift could not be tolerated in a hospital surgical theater. Some applications require that a signal be sent to a centrally located controller to monitor changes in a controlled variable, whereas other applications may require a start/stop action to occur under certain conditions. The point is that careful consideration must be given to the selection and application of sensors to make certain that they are appropriate for the level of control they are required to provide. Important sensor selection considerations are:

1. *Environmental conditions.* Before a sensor is selected, consideration must first be given to the requirements of the environ-

ment being controlled. Each environmental variable being controlled must be evaluated to determine its correct degree of control. The temperature in an office does not have the same criticality as that of a controlled variable in a manufacturing process, for example. Acceptable tolerances for control performance must be established before a particular system is selected.

2. *Operating range of controlled variable.* The span over which a sensor can effectively measure must be matched to the potential range of the environmental variable being controlled. For example, a temperature sensor that can measure temperatures from 0 to 250°F would be inappropriate for controlling an office environment with a control range of 55 to 85°F.
3. *Sensor signal compatibility.* The type of signal produced by the sensor must be compatible with the controller receiving it. A 4- to 20-mA current signal cannot be understood by a controller that measures variable voltages without transducing the signal.
4. *Set point accuracy and repeatability.* A very stringent requirement for accurate control of a set point will also require that the sensor have a high rate of repeatability; that is, the response of the sensor must be the same over time given constant conditions. Not all sensors offer the same level of accuracy or repeatability, and their quality will vary among manufacturers.
5. *Response time.* The response time needed to effect a change in the measured variable is a function of the type of environment being controlled. HVAC systems by their very nature have a significant amount of hysteresis, or lag, because they are built up; that is, because they are composed of many subsystems, the total system will respond very slowly to a change in a control command. Therefore, a fast response in certain circumstances is not desirable because it could cause the control system to "hunt."

Controllers

A *controller* is a device that collects information from a sensor, interprets this information, and sends commands to a controlled device to effect a corrective action. There are five fundamental types of controllers, differentiated by their power source and by how they receive and transmit control signals:

- Pneumatic
- Electric
- Electronic
- Direct digital
- Hybrid systems

Pneumatic controllers, commonly called receiver–controllers, are mechanical devices that respond to changes in air pressure from a pneumatic transmitter and produce a proportional pressure output to a pneumatic control device. They are usually proportional mode controllers, but can include the integral function for reset control. There are two general classifications of pneumatic controllers: bleed type and relay type.

Bleed-type controllers vary their output signal by having control air pressure exhausted, or bled, to the atmosphere from a small air valve that is positioned by the incoming sensor signal. They provide a fast response in one direction because the pressure signal transmitted to the controlled device can be quickly bled off; however, they are slow in the other direction because of the small volume of air that can be transmitted through a restrictor tee. They are also limited by the total volume of the control air supply.

Relay-type controllers actuate a relay that amplifies the control air pressure signal to a controlled device. Because the control air is not exhausted to the atmosphere, this type of controller is less expensive to operate in that it needs fewer units (scfm) of compressed air to operate and has a faster response than a bleed type, as only the pilot line is controlled by the sensor.

Because pneumatic controllers are mechanical devices, they can be configured to operate in both direct- and reverse-acting modes. This feature allows many sensors, controllers, and control devices to be combined to create efficient and cost-effective pneumatic control systems.

Direct-acting controllers increase their output signal in proportion to a rising input signal. For instance, consider a zone sensor and controller modulating a chilled-water valve to maintain space temperature, as shown in Figure 2-12. As the space temperature rises, the sensor signal to the controller rises; this causes the controller to increase its signal to the valve actuator, causing the valve to open and allow cold water to enter the coil. As the space temperature drops, this process reverses itself and continues to alternate to satisfy the set point of the thermostat.

A *reverse-acting controller* decreases its output signal as its input signal increases. (Refer again to Figure 2-12.) The difference is that as space temperature increases the output signal to the valve actuator decreases. This application requires that the valve actuator be arranged in a normally open configuration, which is highly unlikely.

Electric controllers are usually two-position devices. An example is a two-position electric thermostat that closes a relay contact to start an attic fan when the attic temperature exceeds set point, as shown in Figure 2-13. Electronic controllers are proportional and proportional

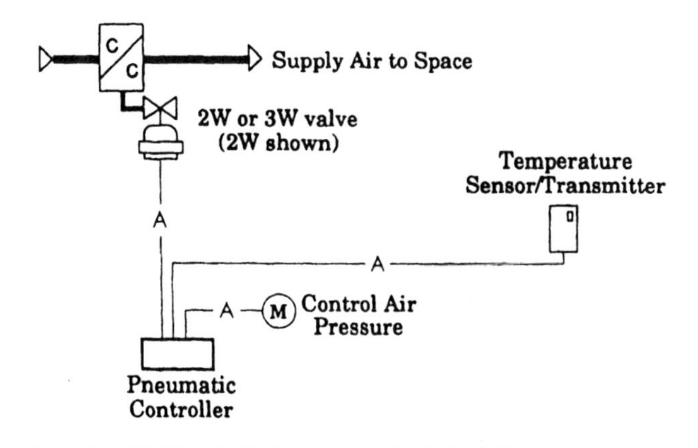

Figure 2-12 Cooling coil valve control (proportional).

plus integral (PI) mode controllers. They have the ability to measure variable electronic input signals and to produce variable output signals. Early electronic controllers were based on the balanced bridge, which is still a popular design in electronic controllers.

Direct digital controllers, like electronic controllers, also have the ability to measure continuously varying input signals from electronic sensors and to produce modulating output signals. However, there is a significant difference as to how this is done. The input signal from the sensor is converted in the controller from analog to digital form. The heart of the controller is a microprocessor that can process multiple loops and, like a computer, accept digital commands from keyboard inputs or digital information from phone lines, perform mathematical

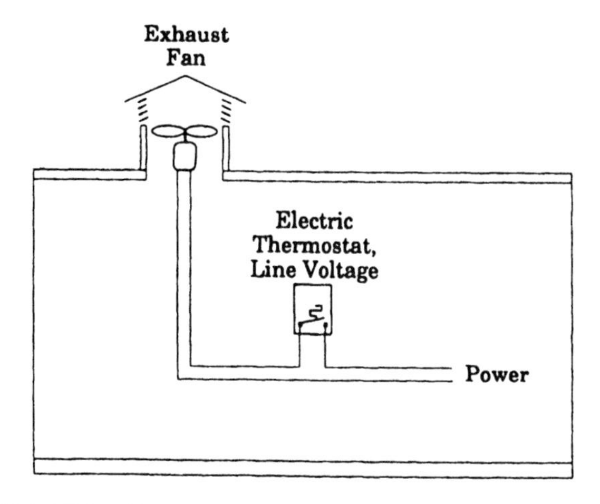

Figure 2-13 Exhaust fan operation control (two-position).

computations, display information on monitor, and perform supervisory and energy management control routines. The output information is reconverted to analog form for transmission to the controlled device. (Refer to Figure 1-1 for a block diagram of this process.)

Hybrid systems are combined configurations of the four types of systems described, created to achieve first-cost economies and optimal system performance. Chapters 4 and 5 explore these combinations for specific HVAC applications.

Single-Purpose Controllers

The thermostat is the single most important device used in modern HVAC control systems. It combines the elements of a sensor and a controller in one compact enclosure and is available in pneumatic, electric, and electronic forms. This section describes common terms and special applications utilizing thermostatic controllers.

Day/night, or *dual-temperature, room thermostats* control space temperature at one set point during the day and at another temperature during the evening. Normally used in colder climates, a night setback thermostat will prevent the building temperature from becoming too cold, thus putting additional demand on the HVAC system to warm up the building in the morning. The switchover from day to night operation is accomplished with a manual or automatic switch. Usually an electric time clock energizes a pilot relay to change over every day/night thermostat in a building.

Combination heat/cool thermostats control valves or dampers that regulate a heating source and a cooling source. This thermostat is toggled between its heating and cooling modes by either a manual or an automatic switch, which usually depends on a measured temperature such as outside air. Submaster thermostats have set points that are adjusted over a predetermined range by the output control signal of a master controller. For instance, a master thermostat measuring outside air temperature can adjust the temperature set point of a boiler hot water supply controller to anticipate the demand for heating in a building. This application is commonly referred to as boiler reset control.

Deadband thermostats have an adjustable differential between heating and cooling, also known as the null band or zero energy band, which is used to prevent simultaneous heating and cooling sequences from occurring. In the neutral position, free cooling can be used if cool outdoor air is available; likewise, a source of free heat can be utilized to satisfy the thermostat set point until it can no longer maintain temperature and mechanical heating or cooling is required. Figure 2-14 graphs the control action of a deadband thermostat.

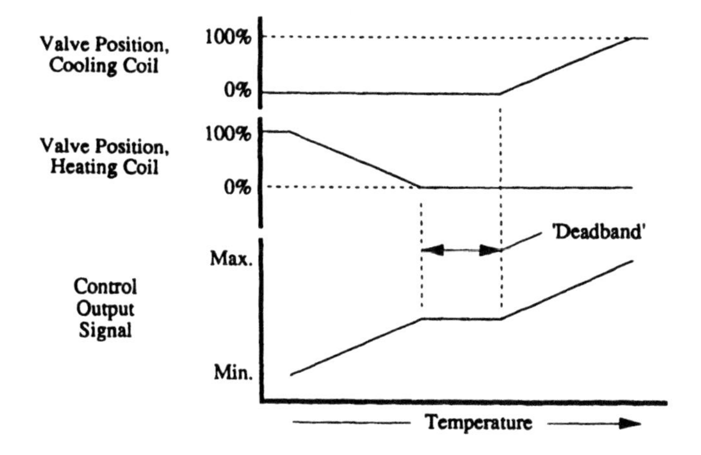

Figure 2-14 Control action of a deadband thermostat.

Controlled Devices

Controlled devices are components that are controlled by mechanical or electronic means. They are often referred to as field devices, meaning that they are located on or near the equipment they operate. Common controlled devices are control valves and control dampers.

Control Valves

Valves are devices that regulate the flow of liquid or gas through a pipe. Many types of valves are used in HVAC applications, such as valves that stop flow (two-position), valves that regulate flow (proportioning), and valves that limit the direction of flow (check valves). The performance rating of a valve under automatic control is based on its flow characteristics as it operates through its full stroke and on the pressure dynamics of the system it is operating within. Although it is the objective of the mechanical engineer to design a system with constant pressure drops across all controlled devices, the very nature of modulating control causes system pressures to fluctuate and therefore to affect the performance characteristics of valves and dampers.

There are three general classifications of automatic flow control valves commonly used in automatic control systems, differentiated by their flow characteristics: equal percentage, linearity, and quick opening.

Equal percentage, or *incremental, valves* modulate through their stroke in percentages of flow that are equal over their entire stroke. In other words, for each increment that the valve opens, the flow through the valve will increase in an increment equal to that increase in the valve position. *Linear valves* are designed to proportion directly the valve position and flowthrough over the range of control. *Quick opening valves* allow maximum flow to be approached at a much lower percentage of stroke than do linear or equal-percentage valves.

Figure 2-9 presents for comparison the flow characteristics of these three types of valves.

Valve Operators

To operate a control valve through its stroke, a controlled device called an *actuator* is needed. A controller positions an actuator by producing a signal that an actuator recognizes and responds to by moving the valve through its stroke. As the controller drives the actuator, the actuator will turn a shaft or stem on the valve, which will position the valve to allow flow to pass through the valve.

There are three common types of valve actuators: electric solenoid, electric motor, and pneumatic operators. *Solenoid actuators* use a magnetic coil that positions a movable plunger. This plunger is connected to a valve stem and moves the valve stem whenever the magnetic coil is energized. Solenoids are usually found in two-position applications; however, they can be modulated by using floating-control signal outputs.

Electric motor actuators utilize a gear train and mechanical linkage to position the valve stem. There are three principal types of electric actuators: bidirectional, spring-return, and reversible.

Bidirectional actuators are two-position; they are either opening or closing. Once started on its stroke, a bidirectional actuator continues until it has reached the maximum position of its stroke.

Spring-return actuators are driven by the controller to a controlled position and held there or modulated. When the control signal is broken or interrupted, the spring returns the valve to its normal or fail-safe position. Reversible actuators provide floating and proportional control. The electric actuator can be driven in both directions and positioned in midstroke.

Pneumatic actuators are spring-loaded mechanical devices that move a valve stem in response to changes in the incoming control air pressure signal. Pneumatic actuators are commonly used for proportional control and are usually equipped with a spring-return mechanism linked in a normally open or normally closed fashion. A normally open actuator returns to the open position when the control signal is interrupted; a normally closed actuator returns to the closed position.

Control Dampers

A *control damper* is basically a valve for air that is made up of a metal frame with blades that are mechanically linked. Control dampers can

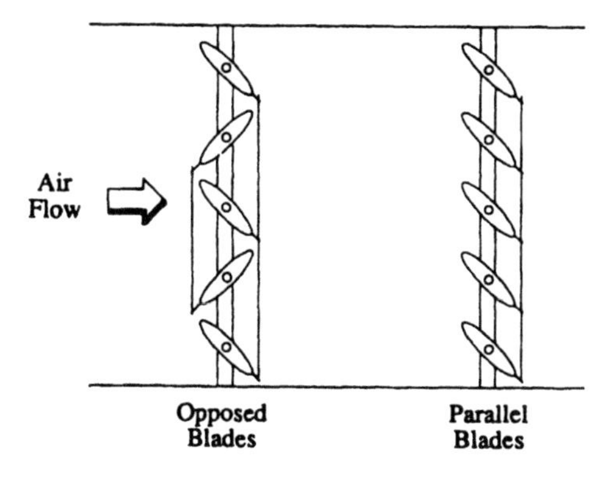

Figure 2-15 Parallel- and opposed-blade dampers.

perform two-position or modulating control, depending on the type of actuator controlling it. There are two types of control damper configurations: parallel blade and opposed blade (see Figure 2-15). Opposed-blade dampers provide more linear control of air than parallel blades and are more suitable for volume control applications, such as duct pressure control where only one damper is used. Parallel-blade dampers offer a lower pressure drop at midstroke and are more suitable for air-mixing applications where the differential pressure drop is low, relatively constant, and not important to the overall sequence of control.

Damper Actuators

Damper actuators are similar to valve actuators and are available in pneumatic or electric form. The primary difference is that the damper actuator is equipped with a rotational shaft designed to be mechanically linked to the shaft of a control damper. The stroke of the damper is measured in degrees of travel across its operating range, which can be easily converted into units of controller output, such as 3 to 6 VDC.

Damper actuators are mounted to a damper frame assembly either in or out of the airstream and are connected to the damper shaft with a mechanical linkage. Large dampers are often mounted in sections so that multiple actuators can be installed that will operate in tandem from one controller output signal.

Positive Positioners

A standard actuator controlling a valve or damper may not respond with enough accuracy to a change in the control signal to satisfy the operating performance requirements of some systems. A *positive positioner* is used to act as a pilot valve that operates with low hysteresis and requires a low volume of control air. Positive positioners provide finite adjustment and repeatable position changes and allow for the adjustment of the control range to provide proper sequencing of multiple controlled devices.

Positive positioners are available in pneumatic form and are used on electric actuators in the form of a "driver," an electric device that "steers" the actuator as it travels through its stroke to maintain accurate control. In pneumatic systems, the combination of relay-type controllers and positive positioners provides superior control performance because they operate from a pilot line using smaller diaphragms and react faster to changes in the controlled variable because of the smaller volume of air in the control line. Pneumatic control manufacturers usually do not recommend bleed-type controls if branch lines are longer than 50 feet or if there is more than one actuator in the control loop.

Peripheral Control Devices

There are many conventional control devices used in conjunction with direct digital controllers to perform specific functions. Although just about any electrical or mechanical device can be monitored or controlled via DDC, a few of the more useful auxiliary control devices are discussed in the following paragraphs.

Electric Auxiliary Control Devices

Transformers are used to provide the proper voltage for electric devices. They can be mounted in panels or located on the equipment being controlled, wherever the conversion of electric power from one voltage to another is required. Relays are frequently used to start and stop devices that have an electric load that is too large for the controller to handle directly. They are also used to create special start/stop sequences and safety interlocks for fail-safe conditions.

Potentiometers are adjustable resistance devices that are used for manual positioning of controlled devices and for remote set point adjustments.

Electric switches are used for two-position manual control of controlled devices and for digital input switching.

Switches can be used wherever human interface to the control system is required or desired.

Step controllers use multiple switches that are activated in sequence to control equipment in several stages. The capacity of a refrigeration machine is modulated by using several states of two-position switching control. They are also used to alternate the operating duty between several machines to equalize wear and tear.

Pneumatic Auxiliary Control Devices

Electric-to-pneumatic (EP) relays and switches are electrically actuated air valves used to operate pneumatic equipment based on the make-or-break of an electric input signal to the relay.

Pneumatic-to-electric (PE) relays and switches are actuated by air pressure from a receiver–controller to make or break an electric circuit.

Switching relays are pneumatic air valves used to divert air from one pneumatic circuit to another for opening or closing air circuits.

Pneumatic switches are two-position devices used to operate pneumatic devices by making or breaking an air circuit.

Peripheral Control Equipment

Air compressors are the source of air used to power pneumatic systems. The air supplied to a pneumatic control system must be clean and dry to prevent contamination or fouling of the air lines and components; therefore, several accessories must be added to an air compressor for this purpose. The air must be filtered to remove contaminant particles from the airstream. A refrigerated air dryer chills the air to precipitate any moisture in the airstream before it is distributed throughout the system.

Transducers are devices that convert signals from one form to another. They are found in systems composed of combinations of electric and pneumatic controls. Transducers can convert a proportional input signal into either a two-position or a proportional output signal. A common example is the PE transducer, which converts a proportional pneumatic signal to a proportional electronic signal, or vice versa. In applications where direct digital controllers are used in conjunction with pneumatic controls, these devices are required to make input and output signals compatible.

Square-root extractors are a special type of transducer that converts an input signal that varies at a squared rate to a linear output. Square-root extractors are commonly used in pressure control systems to provide linear input signals to a controller, such as the airflow measuring station described earlier in this chapter. The principle of operation is based on the squared relationship between volume and pressure. As the volume of a fan system modulates from full flow to half flow, the velocity pressure in the air duct varies from maximum to one-fourth of maximum. Using a common supply-and-return fan system as an example, we know from the fan speed law that if air volume is reduced from 100% to 50% flow, the supply air velocity pressure is reduced from 100% to 25% pressure (0.5^2). If the return fan maximum volume is 50% of the maximum volume of the supply fan, the return air velocity will vary from 25% (0.52) to 6.25% (0.252). A square-root extractor linearizes both the supply and return velocity pressure signals to allow a controller to compare them.

In this chapter we reviewed the basics of temperature control theory and operation. The terminology and control system elements discussed here are vital to a complete understanding of DDC systems. Sensors and controlled devices are, respectively, the inputs and outputs to and from a controller. The relationship between the elements of a control system is the same in conventional control systems as it is in DDC systems.

Chapter 3

Fundamentals of Computer-Based Controls

The computer is the heart of any direct digital control (DDC) system. Although many models of DDC computer systems are currently available, most operate in the same basic way and are similar in design to personal computers (PCs). At the most basic level, all computers consist of hardware and software.

The term *hardware* includes all the components from which a computer is assembled: the central processing unit, memories, input and output terminals, operator display terminal, keypad, and peripheral devices. Peripheral devices are used to link the computer's internal environment to the outside world to allow humans to access stored information and to allow for machine-to-machine communications in a computer network.

Software is a series of instructions that tell the computer how to operate. These instruction sets or commands are packaged together to form programs designed to manipulate data for a specific purpose. There are several echelons of software used in computers, ranging from programs that tell the computer how to handle basic tasks to those that guide the computer through difficult data processing. Computers commonly use several layers of software running concurrently to guide their operation, and these layers are transparent to the computer operator. The nature and functional aspects of general-purpose and specific-function software are considered in this chapter.

Microcomputer Fundamentals

Data Representation

The numbers 0 and 1 are the primary language elements of all computers. By grouping these two digits in different sets and combinations, virtually any kind of information, in the form of either letters or numbers, can be presented to the computer in a format that it can understand. The most common numbering systems used in computers are the *binary system,* or base 2, and the *hexadecimal system,* or base 16. Any system of numbers using a fixed base value can be used, such as the octal system of base 8 or decimal system of base 10; however, it is more efficient to use a system with a base that is a power of 2, to allow whole number exponential expansion to occur. The binary system of arithmetic is usually used as a framework to convert all data and instructions into logical series of binary digits, or bits, representing combinations of the numbers 0 and 1. A 0 represents a low-voltage value, 1 a high-voltage value. A computer communicates information both inside and outside of its processing environment in the form of words that are groups of numbers. A computer word is the most fundamental unit of information used by a computer, and can represent a letter, a symbol, a number, or a specific instruction for the computer to follow.

Computer words consist of groups of bits, commonly referred to as *bytes.* Words are used to express both numeric and text characters. As with any language, words are not useful unless they are formed into sentences to create communication; likewise, bytes must be grouped together in strings to form program sentences that the computer can understand. A *string* is a series of consecutive bytes that are treated as a single entity by the computer. A single byte is comprised of 8 bits and can represent values from 0 to 256 by using combinations of 0s and 1s over a parallel eight-digit range, which is expressed as (2 × 108). A 2-byte, or 16-bit, word represents numbers up to 65,536 (2 × 1016). In this manner, more than 65,000 memory addresses, program instructions, or calculations can be communicated to the computer within the space of one word. The speed of the semiconductor determines the number of words that can be processed in a given time frame.

As the technology of semiconductor manufacturing evolved, the word length capability of computer chips grew. In 1971, Intel Corporation released a 4-bit microprocessor architecture that was the first of its type; soon afterward, this chip was replaced by 8-bit microprocessors from many manufacturers. As the density of semiconductor component materials increased, the power of these processors has increased exponentially from 4-bit to today's 64-bit data manipulation, all within the framework of a single silicon chip. The longer the word length, the larger the numbers the computer can manipulate. In addition to these improvements in capability, the electric power required to drive later generation chips was greatly reduced, allowing them to perform tasks equivalent to those of thousands of transistors without requiring space to dissipate collected heat. Hence, as chips become ever more powerful, they also decrease in physical size. It is interesting to note that in the recent past the primary difference between mainframe computers, minicomputers, and microcomputers was indexed to the word length that their processors read. Traditionally, mainframes employ 32-bit words or larger; minicomputers 16- or 32-bit words; and microcomputers 4-, 8-, 16-, or 32-bit words. Using word lengths to differentiate system types is no longer valid, because there are many microcomputers that have the word length capabilities of microcomputers or mainframe machines as a result of innovations in microprocessor technology. Computer programs consist of numeric strings that form the code that guides the computer in every function it performs.

Code is a term used to describe the direct relationship between a character and the binary number indexed to this character. As you can imagine, juggling large groups of bits to write program code can be cumbersome and confusing. Therefore, a form of binary shorthand, called hexadecimal, or base 16, notation, was developed, which is a combination of the terms "hexa," meaning six, and "decimal," meaning ten. It is quite simple to use because it is based on both base 2 and standard base 10 mathematics. The hexadecimal system uses 16 digits that can be sequenced to express millions of different quantities. Because our traditional numbering system contains only the 10 digits 0 through 9, the letters A through F are used to represent the numbers 10 through 15. With hex notation, one hex digit is equal to 4 bits. Therefore, if we consider the values 0 and 1 as a grouped pair, there are four possible combinations of these two values, ranging from 00 to 11; by lumping these two-digit combinations into groups of four in hex, we can achieve 24 or 16 different combinations of bits ranging from 0000 to 1111. Because a byte is comprised of 8 bits, hex can conveniently express a byte with two hex digits. Obviously, this is much easier to deal with than traditional bit notation. Hexadecimal system values and their decimal equivalents are shown in Table 3-1.

Text characters, such as the letters A, B, C, and D, each have distinct and individual bit patterns, or codes, that identify them from all others. Each character occupies a single 8-bit byte; therefore, there are 28 or

Table 3-1 Hexadecimal Value Table

Hex	Dec	Hex	Dec	Hex	Dec	Hex	Dec	Hex	Dec
1	1	10	16	100	256	1000	4096	10,000	65,536
2	2	20	32	200	512	2000	8192	20,000	131,072
3	3	30	48	300	768	3000	12,288	30,000	196,608
4	4	40	64	400	1024	4000	16,384	40,000	262,144
5	5	50	80	500	1280	5000	20,480	50,000	327,680
6	6	60	96	600	1536	6000	24,576	60,000	393,216
7	7	70	112	700	1792	7000	28,672	70,000	458,752
8	8	80	128	800	2048	8000	32,768	80,000	524,288
9	9	90	144	900	2304	9000	36,864	90,000	589,824
A	10	A0	160	A00	2560	A000	49,960	A0000	655,360
B	11	B0	176	B00	2816	B000	45,056	B0000	720,896
C	12	C0	192	C00	3072	C000	49,152	C0000	768,432
D	13	D0	208	D00	3328	D000	53,248	D0000	851,968
E	14	E0	224	E00	3584	E000	57,344	E0000	917,504
F	15	F0	240	F00	3840	F000	61,440	F0000	983,040

256 different characters that can be represented in byte form in an 8-bit format. The American National Standards Institute (ANSI) developed the American Standard Code for Information Interchange (ASCII—pronounced "as-key") to provide a means of standardizing the symbols used to represent letters and numbers in computing environments. Although other standards have been developed, the ASCII standard is the most common code in use today. The ASCII code uses a 7-bit code string; therefore, it is limited to 27 or 128 different symbols. These symbols, called the *character set,* and their code equivalents are shown in Table 3-2.

The purpose of presenting hexadecimal here is that computer code is commonly expressed in hex format. The more you are exposed to computer systems and the software programs they run on, the more likely you are to see hex-based code. This is how computers store data; only after these numeric data are translated into our language can they be understood and manipulated by programmers and users.

Elements of Microprocessor Controllers

There are five essential elements of a microprocessor controller:

- Central processing unit (CPU)
- Program memory

Table 3-2 PC Character Set with Decimal & Hex Codes

Dec	Hex	Char. Set	Dec	Hex	Char. Set	Dec	Hex	Char. Set
48	30	0	74	4A	J	100	64	d
49	31	1	75	4B	K	101	65	e
50	32	2	76	4C	L	102	66	f
51	33	3	77	4D	M	103	67	g
52	34	4	78	4E	N	104	68	h
53	35	5	79	4F	O	105	69	i
54	36	6	80	50	P	106	6A	j
55	37	7	81	51	Q	107	6B	k
56	38	8	82	52	R	108	6C	l
57	39	9	83	53	S	109	6D	m
58	3A	:	84	54	T	110	6E	n
59	3B	;	85	55	U	111	6F	o
60	3C	<	86	56	V	112	70	p
61	3D	=	87	57	W	113	71	q
62	3E	>	88	58	X	114	72	r
63	3F	?	89	59	Y	115	73	s
64	40	@	90	5A	Z	116	74	t
65	41	A	91	5B	[	117	75	u
66	42	B	92	5C	\	118	76	v
67	43	C	93	5D	]	119	77	w
68	44	D	94	5E	^	120	78	x
69	45	E	95	5F	–	121	79	y
70	46	F	96	60		122	7A	z
71	47	G	97	61	a			
72	48	H	98	62	b			
73	49	I	99	63	c			

- Data memory
- Input/output interface
- Clock circuits

The *CPU* is a microprocessor, which is a logic device that performs arithmetic, logic, and control operations. It has the ability to carry out instructions through the computer and is contained on a chip of silicon material that is one-tenth of a square inch large. A semiconductor chip is the brain of the microprocessor and is comprised of an arithmetic logic unit (ALU) and a control unit combined onto a single chip. One of the most important circuits in a microprocessor, the ALU performs arithmetic and logic operations on information that is sent to it. The control unit is responsible for collecting program instructions from the computer memory section in proper sequence and feeding this information to the ALU for processing. The ALU then sends processed instructions to the output section, which converts the results into a signal that can be understood by the peripheral equipment the computer is controlling. All communication in a computer is done with binary state representations of ordinary numbers, which will be explained shortly.

A block diagram of a microcomputer is shown in Figure 3-1. It is important to note the relationship between the microprocessor and the microcomputer; although these two terms are frequently used interchangeably, they are distinct architectures within the framework of the system. The term *microprocessor* refers to the address and register blocks within the digital chip; *microcomputer* refers to the processor chips, memory chips, I/O ports, and clock circuits that combine to create a controller.

For a microprocessor to operate, a common communication channel is needed throughout the computer to connect the CPU with the other devices in its circuitry. A group of conductors known as a *bus* carries information between the elements of the microcomputer. Think of a bus as a party line telephone system that allows simultaneous communication to take place between several devices connected to it. Each

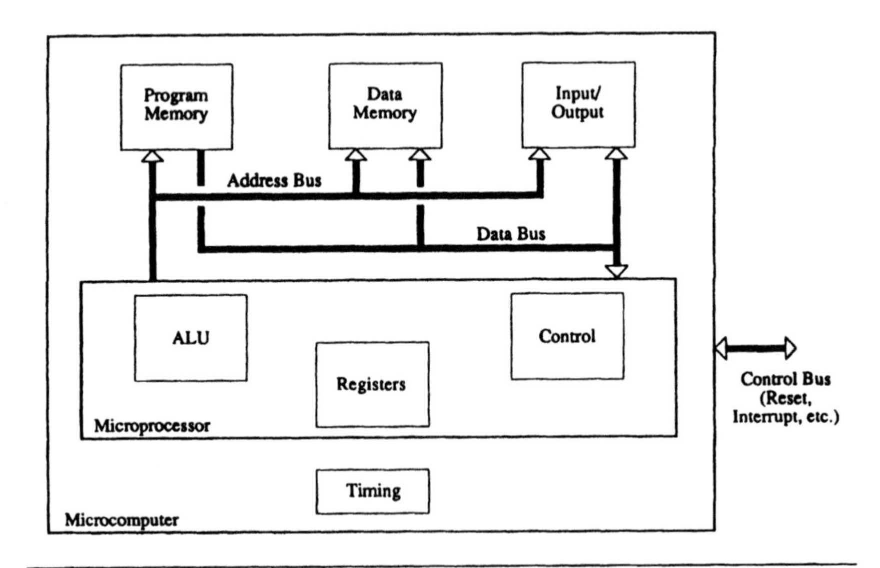

Figure 3-1 Relationship between the microprocessor and microcomputer.

device on the bus has an address to allow the CPU to call specifically on each device to perform a service when it is needed.

There are three distinct buses in the microcomputer architecture: a data bus, an address bus, and a control bus.

A *data bus* carries words between the ALU and the memory and usually has one conductor for each data bit. Therefore, a 16-bit CPU has a 16-line data bus.

The control unit has an *address bus* that it uses to collect data and instructions from memory for delivery to the ALU for processing.

The *control bus* has access to every device in the computer. It controls the execution of program sequences by controlling the flow of instructions and data throughout the computer during operation.

Program memory stores information on how the microcomputer is supposed to operate. This is where the computer receives commands to do assigned tasks. This information is in the form of instructions or programs that the computer must follow in order to operate. Programs tell the computer what to do and how to manipulate the input data it receives. It is where all of the activity of the system takes place. The size of the memory puts a practical limit on the amount of work that the computer can perform.

Data memory is the equivalent of a filing cabinet or library for the computer. Data are stored on a disk when not being used inside the computer memory. Disk storage can be internal or external to the computer.

Input/output interface is the means by which the computer takes in and sends out information. It is the interface between the computer and the outside world. Input can be in the form of incoming control signals or information that is typed onto a keyboard. Ports are used to connect the computer to peripheral devices for both incoming and outgoing data exchanges.

A *serial port* is used for two-way communication between the computer and its peripherals, and it is usually the point of connection to a telephone line or local area network. Serial ports are general-purpose ports that can be used to connect any peripheral to the computer system. To connect a serial port to a standard telephone system, we need a method of translating computer signals into telephone signals. This is provided by a *modem,* which is an external device connected between the serial port and the incoming telephone line. An internal modem combines a serial port and a modem onto a single circuit board inside the computer enclosure.

Parallel ports are used for special one-way communication between the computer and a printer.

The *clock circuit* is the master timer of the system. It acts like a metronome in that it sets the pace for the work performed by the computer.

Electronic Memories

The memory of a computer is analogous to a large desktop within the computer; it is a working space where the computer can spread out information it is working on, such as program instructions and input and output data. This information is stored in the form of coded bit patterns, each with a unique address so the computer can easily find it on its "desktop."

Computer memory is organized into three basic categories based on the nature of the information being stored: permanent, changeable, and temporary. These memories are arranged in layers inside the computer to allow the sharing of information to occur.

Permanent Memory

Permanent memory holds programs that supervise the computer in the performance of its basic operations. *Read-only memory* (ROM) contains unchangeable program instructions that are installed when a computer is assembled, which can never be altered or reprogrammed after they are "burned-in." Because this software is installed permanently, it is called firmware. ROM is considered to be a low-level memory because the operation of the programs in ROM is invisible to the computer operator. The computer accesses ROM memory internally as needed to run its programs. Although the code in ROM is permanently installed, it does require a constant source of power to remember the information it stores. Fortunately, the power required to support a microchip is very low; usually, a low-voltage rechargeable battery is installed in the computer for this purpose.

A special set of programs is stored in ROM that connects the hardware devices of the computer to the software programs that serve the needs of the user. *Basic input/output services* (BIOS) are software programs that bridge higher level software to hardware by using ports. A port is the point of connection on a hardware device. By a communication bus, each piece of hardware in a computer system can be linked on a "party line" and given an address that can be called whenever hardware services are required. This bus is reserved exclusively for use by the BIOS and is part of the circuitry of the computer system.

Another memory technology that has recently been used in DDC controllers is known as *nonvolatile ROM* (NVROM). NVROM is a permanent memory that does not require a source of power to maintain the contents of its memory. Whereas a ROM memory requires a battery to keep the memory in check, NVROM can be applied to systems that may experience power interruptions. Because the power supplies in industrial environments are prone to sudden voltage fluctuations or "spikes," NVROM chips in microprocessor controllers offer a significant benefit over traditional ROM chips.

Changeable Memory

The next echelon of computer memory is called *changeable memory,* which is a derivation of ROM that allows changes to firmware to take place without the need to replace the memory chips. *Erasable programmable read-only memory* (EPROM) is used in dedicated computing applications where there is a potential need for reprogramming in the future. By introducing a slight electrical charge to the EPROM chip, the memory is erased and can then be reprogrammed. EPROM memories are often used in such familiar equipment as "smart" thermostats and variable-frequency-drive controllers.

Temporary Memory

Temporary memory is used to store information that is being used only when the computer is performing an operation. *Random-access memory* (RAM) is a volatile form of memory that allows the computer to read and write information of a temporary nature whenever it needs to. Again, think of computer memory as workspace for the computer. Temporary data that the computer needs to perform a routine are placed in temporary memory. Under the direction of the BIOS, the computer can read from and write on the memory at any time. For instance, consider a simple software algorithm that resets the position of a hot water valve based on outside air temperature. The current air temperature is a temporary value, as it will likely change within moments. The computer writes the air temperature onto the memory, and then reads this value to calculate the value of the output signal it must send to the valve actuator. The next time the computer records the air temperature value, it will write the new value over the previous value, thereby erasing the previous value. This process of reading and writing will continue for as long as the computer is running this routine. Once the computer is turned off, it "forgets" everything previously written in RAM (hence, the term "volatile") and will reuse this space to store other temporary data the next time it is turned on.

Peripheral Devices

Computer-based control systems rely on peripheral devices to exchange information between the CPU and its external environment.

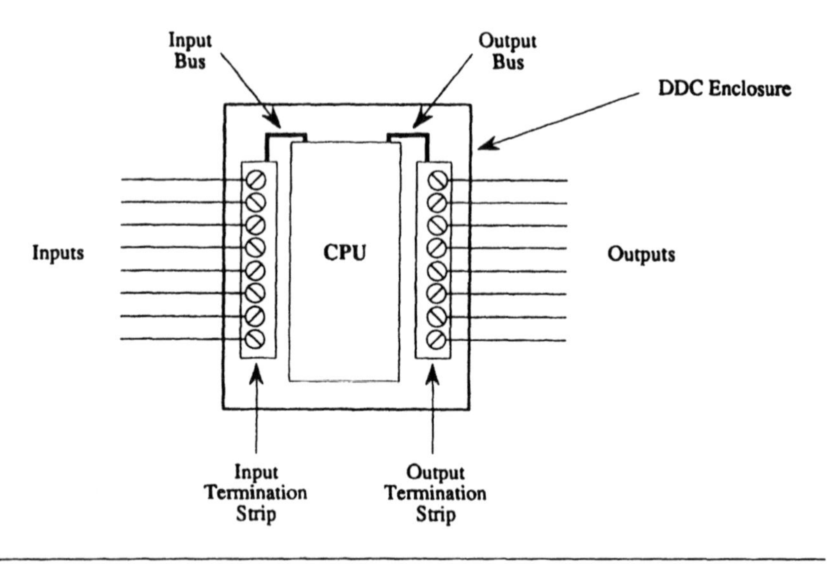

Figure 3-2 CPU input/output connection.

An *input/output device* is any device that provides a point of connection between incoming information to the computer and outgoing signals from the computer to the devices it is controlling. Termination strips are commonly used for this purpose. In Figure 3-2, a DDC enclosure is shown. On the left side of the panel is a termination strip for input signal wires. This termination strip is connected to the CPU by a wire bus. The right side of the panel holds an output termination strip, also connected to the CPU by a bus, which is the point of connection for wiring that goes out to devices under control.

There are three primary types of I/O devices: field interface panels, data access devices for man–machine interface, and mass storage devices.

Field interface panels, also referred to as *field interface devices* (FIDs), are nonintelligent enclosures that hold termination strips. Their purpose is to provide an orderly point of confluence for many input and output signal wires for a control system. Typically, FIDs are located near groups of devices under control to reduce the amount of field wiring and to allow a single wire bus to be run between separate FID panels back to the CPU.

Data access devices provide a means for human access to information on the operation of the system and allow changes to the system operating parameters to be made. Monitors and printers connected to the computer through data ports are used to provide visual and printed information to an operator for these purposes. More specific information will be given on the characteristics of computer graphics later in this chapter.

Mass storage devices are used to supplement the storage available within the computer memory. ROM and RAM are sufficient to support current operations, but they are not intended for storing historical data or program information that is not currently being run on the CPU. Storage of such information is accomplished with two types of storage media: permanent and removable. The permanent medium is the hard disk drive, which is installed inside the computer enclosure or located outside of the enclosure and connected to the computer by a special cable. The removable medium is the floppy disk, which requires a floppy disk drive either as an integral part of the computer or as an external device similar to the hard drive. The primary difference between the two media is storage area. A floppy disk contains a single disk of magnetic media that can hold up to 1.4 million bytes of information; a hard drive contains a magnetic disk of larger diameter and magnetic density that can hold hundreds of millions of bytes of information. The benefit of floppy disks is that they can be easily removed and transported from machine to machine. The advantage of storing data on a hard drive is that the computer and its operator can access a lot of information very quickly because it is held in memory awaiting a read/write command from the CPU.

Digital information is recorded on magnetically sensitive material, usually iron oxide, which covers a round plastic disk. The surface of this medium is seen by the computer as an array of dots, each representing a bit of information, which can be assigned the value 0 or 1. For the computer to be able to read information from a disk, the disk must be formatted in such a fashion as to ensure that the arrangement of bits is the same for each disk it is supposed to read.

Disks are segmented into tracks and sectors. Tracks are the concentric circles of which the disk is made. The total surface of the disk is divided into these tracks, starting from the outside of the disk all the way to the center or innermost track of the disk. The quantity of tracks on a disk varies with the type of disk being used. The terms *single density, double density,* and *quad density,* when used to describe disks, refer to the number of tracks the disk contains. The more tracks, the more information the disk can store.

The term *sector* refers to logical divisions made to the diameter of a track. Circular tracks are usually divided into 10 to 20 tracks, depending on the density of the disk and the way it has been formatted.

The arrangement of tracks and sectors affects the speed with which the computer can access information via the read/write head of the disk drive. The computer accesses the data on a disk by performing a read/write command. Think of a disk as a phonograph record, and the read/write head of the disk drive as the needle. The needle moves across the disk searching for specific bits of data. The rate at which the head can access information depends on how fast the disk rotates. A floppy drive rotates at 300 rpm, a hard disk at 3600 rpm. Disk drives are evaluated based on the speed with which they can read and write information; obviously, the faster that a drive can access information, the quicker the computer system can perform calculations requiring stored information.

Organization of Computer-Based Control Systems

Computer System Architecture

Computerized control systems comprise three architectural hierarchies that must be understood before we can begin to evaluate computerized control systems:

- System architecture
- Microcomputer architecture
- Microprocessor architecture

System architecture, the broadest level, refers to the integration of computer hardware and software to create a system that performs a specific set of functions. Figure 3-3 illustrates a complete system architecture. The components of the system are arranged in blocks to show how information is organized within the system.

Microcomputer architecture refers to the specific group of components that combine to create a discrete control unit, such as a direct digital controller, which is but one of many possible blocks in a system.

Microprocessor architecture consists of the internal circuitry of a microprocessor, which in turn is an element in the microcomputer architecture, which in turn is an element in the system architecture. The arrangement of the components inside of the DDC panel is its architecture.

To achieve truly integrated control, we need access to information at the deepest levels of the building control system. The more abundant the information provided to the controller, the more efficiently the system can perform. Before the advent of computers, hard-wired control systems were not cost-effective in linking groups of control blocks, and therefore pneumatic and analog electronic control systems had few interrelated control routines between blocks. The prohibitive cost of building customized control panels allowed highly integrated control systems only in large installations.

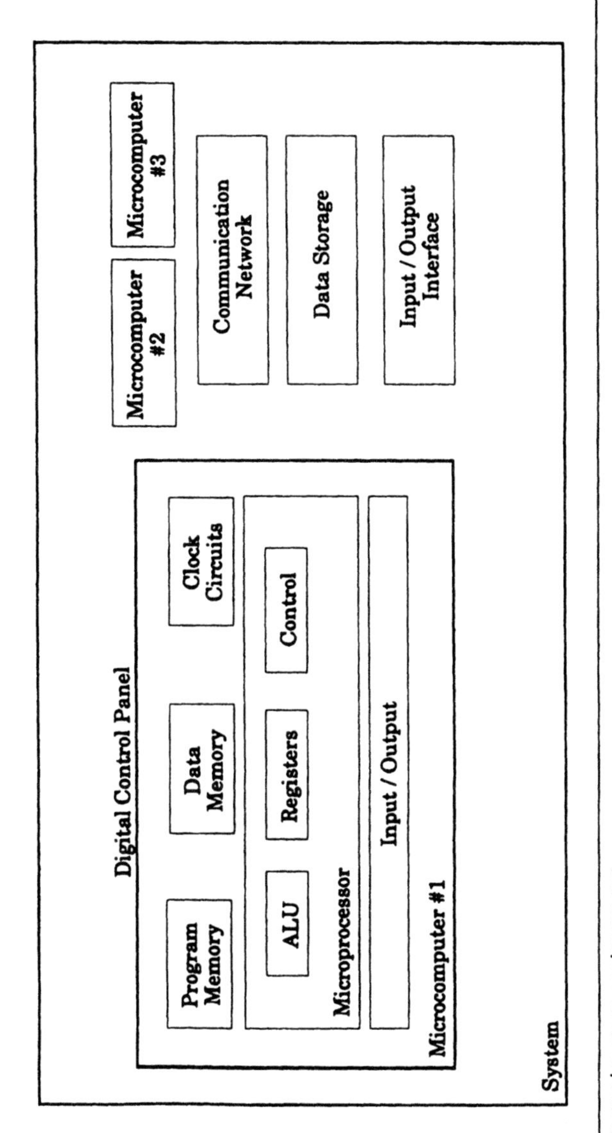

Figure 3-3 Control system environment.

When mainframe computer technology became available to the mass market in the form of PCs, a cost-effective way to achieve highly integrated control was created. Control blocks could now be linked in a hierarchy so that any controller in a system could know what every other controller was doing.

Discrete systems, as we have seen, are modular in nature. Component parts are combined to create control systems that are adaptable to a diverse range of control applications. Similarly, the components that make up building automation systems are modular and work together through a communications hierarchy to perform integrated control routines that are distributed to the farthest reaches of the building HVAC system.

A simple way to understand hierarchical control systems is to relate them to the structure of a business enterprise. Starting at the top, we have a director, or host computer, that watches over the entire control system. It collects information on the operation of the HVAC system, looks for deviations from standard control behavior, monitors changes in the controlled environment that are indicative of future problems, and, when necessary, overrides the commands of subordinate controllers responsible for unitary control. A host computer is typically a PC utilizing custom software that allows an uncomplicated interface between the person operating the building and the control system. Many available software programs offer graphic depictions of real control situations, which are beneficial to the building operator when a human decision interface is required, such as in an alarm condition.

Below the host computer is the network controller, which is responsible for managing the communications traffic between individual DDC controllers. The network controller closely monitors the control routines performed at the field level and facilitates the sharing of information between field panels by collecting information and then distributing it to the field controllers on demand. The network controller communicates vertically by means of a communication bus or *local area network* (LAN) to both the host computer above it and to the field control panels below it. This configuration is commonly referred to as *master–slave,* where multiple slave controllers are subordinate to one master controller. The primary difference between the two is that a master controller has decision-making capabilities whereas a slave can only gather information and return it to the master controller, where it is then processed. Slave panels are referred to as *data acquisition panels* (DAPs).

At the last tier in the hierarchy are the agents of the control system that perform the brunt of the work of controlling field equipment. At this level two types of controllers are used, which are either intelligent or nonintelligent. Nonintelligent controllers have no means to measure, process, or control; they are the point of collection for control signals that must be sent to a network controller for processing. Intelligent controllers contain microprocessors that can run control programs independent of the network controller. Intelligent controllers obviously have the advantage of speed in executing control routines because no time is lost in sending and receiving control actions back from a network controller.

Computerized Control System Architectures

Building automation systems are categorized by the functions they perform. They range from systems that simply monitor building conditions to systems that monitor environments, control equipment, and apply artificial intelligence to execute complex control routines in anticipation of future events. Three systems used are the energy management system, energy management and control system, and facilities management and control system.

The *energy management system* (EMS) encompasses control devices and systems ranging from simple two-position timer controllers to sophisticated digital computers. Just about any product that can reduce energy consumption is being touted as an energy management system. The main difference between true EMS and glorified time clocks is the ability of the EMS to receive and transmit analog input and output signals. Energy cannot be effectively managed without closed-loop feedback; all of the control strategies discussed in Chapter 6 depend on accurate feedback. For instance, an outside air temperature signal tells the EMS when to fire the boilers or to what degree to modulate a hot water mixing valve; without an analog feedback signal, the system would be operating in an open-loop fashion. EMS systems are commonly used for monitoring energy demand, indicating status, and producing alarm responses to status changes in control conditions.

Energy management and control systems (EMCS) perform control routines, monitor building conditions, measure energy consumption, and execute control actions to manage the variables of an HVAC system in accordance with strict instructions given to the system by the system programmer and building operator. They are supervisory in nature and are meant to assist building operators by collecting and evaluating information.

The *facilities management and control system* (FMCS) and *building management system* (BMS) grew out of the integration of HVAC control, lighting control, fire and life safety control, and security control systems by using a common computer to monitor and override the operation of these individual systems. FMCSs provide monitoring and control functions as well as support communication between the HVAC control and other systems within the building. This communication is transmitted digitally, usually by relay contact closures to alert other systems to a problem that has emerged.

Centralized Control Systems

A centralized control system starts with a general-purpose computer, usually a mini-mainframe that is capable of handling a lot of data at very high speeds. All of the programs that operate the HVAC controlled devices reside in the central computer. The central computer console has an input keyboard and monitor and usually includes printers, data logging devices, intercoms for operator communication to remote points in the system, and auxiliary monitors for system interface. Figure 3-4 illustrates the components in a centralized system.

Centralized control systems are supervisory in nature, in that they depend on communication with remote control panels for information on the operation of the system. Distributed throughout the building are data acquisition panels used to link the sensors and field devices in the system to the central computer. The most distinguishing characteristic of central systems is that all control decision-making occurs at the central computer. Local data acquisition panels serve no other purpose than to collect and distribute information to the devices being controlled.

Central systems were first used in medium to large buildings as a means to collect data on the operation of the HVAC system while maintaining control over an entire facility from one computer.

The primary benefits of central control systems are their ability to handle many control loops simultaneously and to provide individually customized programming for each loop. Other benefits of this type of system include operating costs that are lower than for a conventional control system, improved control system performance, and the ability to integrate monitoring and control routines with life safety and security management systems and to collect and organize valuable information used to maintain and manage large buildings.

There are, however, disadvantages to central systems in HVAC applications that must be considered. Because they use mini-mainframe computing machines, they are difficult for some building operators to understand and troubleshoot unless they possess a knowledge of computers and microelectronics. Without intensive operator training, lack of system knowledge unfortunately leads to nonutilization of the system and eventual sabotage. Turnover in building operations personnel requires an ongoing commitment to training operators how to use the system; this training is costly, and is usually omitted. When a problem erupts within the control system, the central computer is

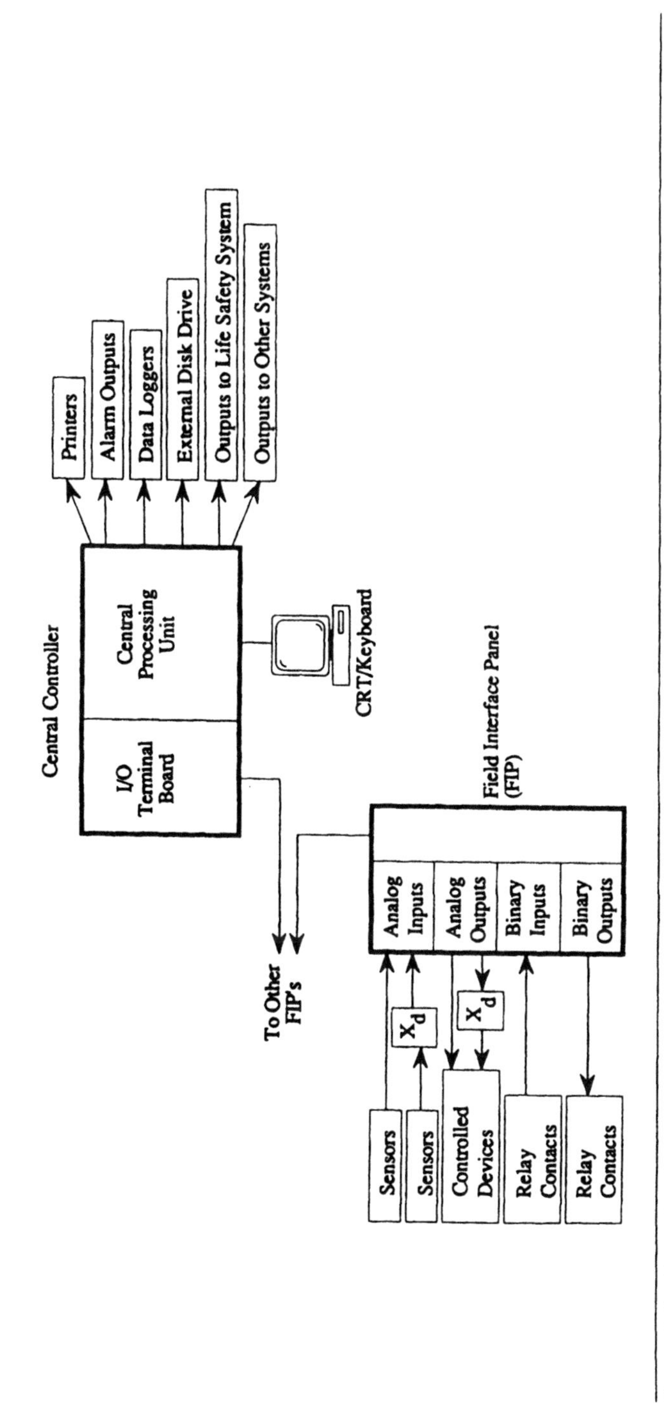

Figure 3-4 Centralized control system architecture.

generally mechanically or electrically bypassed or replaced with redundant conventional controls.

Another serious concern with central systems is the potential for a central computer failure. The operation of the entire control system depends on a single machine; if the central computer crashes, control of the entire control system is lost. Most HVAC control routines are too complicated to be controlled by manual backups; therefore, control system engineers usually incorporate redundant hard-wired conventional controls into their design strategies to avoid catastrophe in the event of a failure. This redundancy increases the first cost of the system.

A final concern is the source of continued support for the system software. Customized software written for a specific building HVAC system will require periodic maintenance as well as corrections to aberrations that often occur.

Distributed Control Systems

In distributed control systems, microprocessor control panels are located near the equipment being controlled and contain all of the intelligence needed to control its assigned equipment fully without intervention by a central computer. Communication takes place between the individual control panels over a LAN, which is connected to a central computer (see Figure 3-5). Although the local control panels act independently of the central system, they benefit from being able to access information from other local control panels in the system.

The true measure of a distributed control system is whether the local controller can perform full HVAC control routines without intervention by a central computer. A truly distributed control system can operate during a loss of communication with the central computer without any degradation in its performance. No redundant electronic control is required as a backup to the DDC system. For this reason, distributed system architectures are considered the most reliable form of computerized control for HVAC systems.

Distributed systems combine the advantages of local and central control methods to provide fail-safe, modular, and cost-effective control for HVAC systems. The most significant disadvantage to distributed control systems is the lack of compatibility between systems of different manufacture. Unlike the personal computer industry, which utilizes a chosen standard (MS-DOS), control manufacturers use proprietary communication methods to prohibit the interface of their systems with other control systems. In the recent past, unless a sole source of DDC equipment was used for a building HVAC system, it was unlikely that all control devices would be able to share information on the same

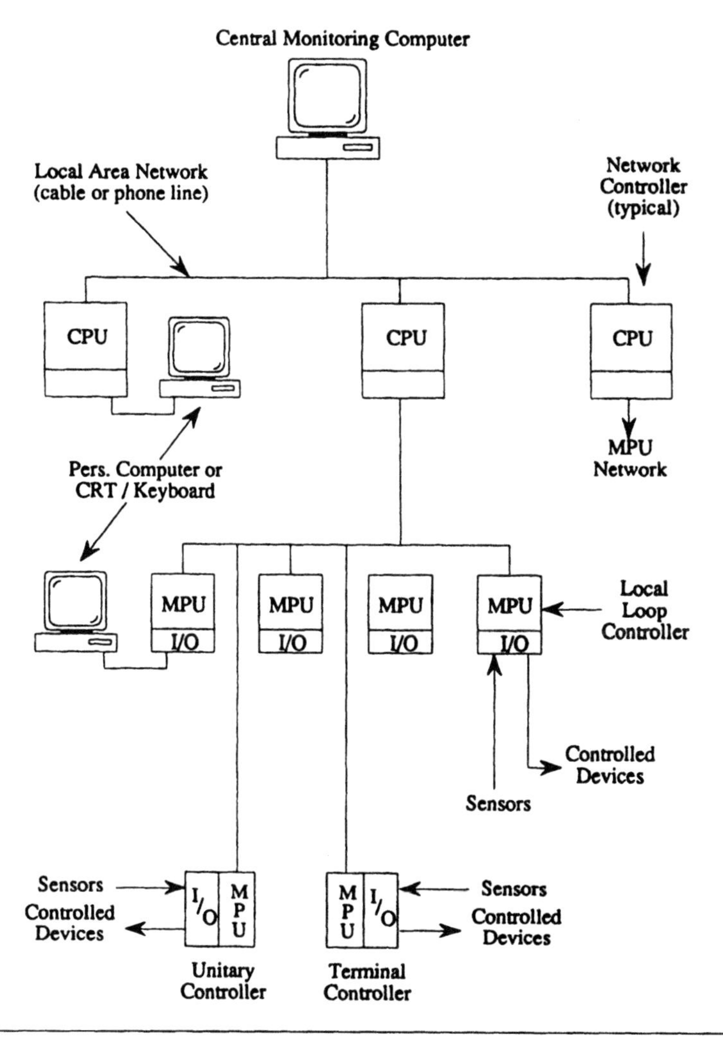

Figure 3-5 Distributed control system architecture.

communication network. Pressure is being placed on the controls industry by building owners and operators to develop a common interface that will make all brands of control systems compatible. In Chapter 5 advancements in this area are explored.

Stand-Alone Control Systems

The term *stand-alone* is used to describe any controller that possesses the intelligence to perform control routines without intervention by another controller. The primary element of distributed control systems is the ability of the digital controller to perform stand-alone local-loop control.

Local-loop, stand-alone controllers were designed to optimize the operation of environmental control equipment and are commonly used for single-zone, multizone, and variable-air-volume air-handling systems and boiler control systems. Stand-alone controllers can sense, remember, decide, and act independently; they do not rely on a superior controller in the system hierarchy for operating instructions; they can, however, accept commands to override current operating routines from a central computer. Common types of local control applications are examined in Chapter 6.

Significant recent developments in the microprocessor industry have made it possible to distribute stand-alone control to the lowest levels in an HVAC system, known as the unit level. Equipment such as small air handlers and variable-air-volume (VAV) terminals can now be controlled by dedicated controllers that contain intelligence. Digital unitary controllers and digital terminal unit controllers are stand-alone digital controllers that operate independently while communicating important operating information over a communication network to other intelligent devices in the HVAC control system (see Figure 3-6). This sharing of information makes it possible to identify energy-saving opportunities and to provide superior feedback to building engi-

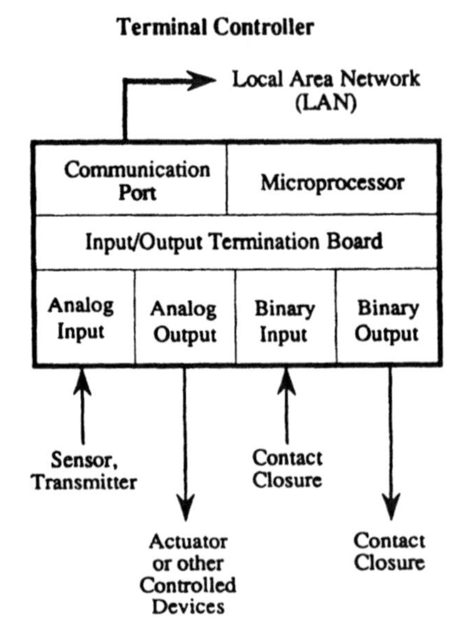

Figure 3-6 Terminal controller architecture.

neers on the operation of the HVAC system; however, a means of communication control is necessary.

Host Computers

A host computer is used to provide centralized supervision and control of a distributed control system. The host acts as a portal to the control system by allowing the operator to oversee the operation of the control system without interfering with the execution of control programs. Host computers are usually PCs equipped with special-purpose software that captures data from the control system in real time to give system operators pertinent information without requiring an extensive data search of the control system. Modern host systems utilize sophisticated graphics software to provide visual images of the HVAC system under control while providing real-time data on the graphic screen to indicate the current state of the system being controlled. Graphics-based host systems provide a friendly interface for system operators who are unfamiliar with computer-based control systems, and they have greatly contributed to the acceptance of computer-based distributed control systems by the building industry.

Computer Communication Systems

Microprocessor-based control systems utilizing more than one intelligent device require a means of sharing information to facilitate the transfer of data between components of the system, to allow remote monitoring of the performance of the system, and to allow for data retrieval and analysis without interrupting the operation of the control system. The transfer of binary data between devices in the system is accomplished on a data highway or local area network.

Local Area Networks

A *local area network* (LAN) is a communications network that allows the interchange of information between individual DDC panels. The IEEE 802 Standards Committee defines a LAN as "... a data communications system which allows a number of independent devices to communicate with each other." The term "local" can mean communication within a single building or, via telephone lines, communication with systems in remote locations.

A LAN comprises three elements: nodes, servers, and connections between the nodes. *Nodes* are intelligent devices on the LAN that support themselves as well as other connections on the LAN. Examples of nodes range from a central host computer to an individual DDC controller. A LAN is transparent to the operation of the individual nodes and allows for totally independent control of each individual system connected to it. A *server* is a software utility that provides services to the nodes on the LAN. Through the server, individual DDC control units on the LAN can receive instructions from control units at higher or lower levels in the system architecture. The server in most distributed control systems is provided in firmware installed in each node on the network.

Frequently referred to as communication cards, ROM microchips containing network program commands tell the local-loop controller how to communicate on the network. Information travels over the LAN at both low and high speeds; obviously, the faster the information can travel, the faster the system can respond to changes in the controlled environment. Unlike common carriers, LANs are not subject to speed limitations and can be designed to transmit information at speeds ranging from 75 bits per second to more than 140 million bits per second. This rate of communication speed is known as the baud rate. Speed ranges common to DDC system LANs are between 300 baud and 64,000 baud, or 64 kbaud.

Types of Networks

There are two primary types of networking systems. The first type is a single-path system known as baseband; the second is a radiofrequency method known as broadband. The difference between these two methods lies in the way in which binary signals are sent over the network wiring system.

Baseband systems utilize a single cable as the path for data to travel over, which is analogous to a one-lane highway wide enough for only one vehicle at a time to travel without a collision. To prevent the data from colliding and to help data travel in an orderly fashion, a network traffic control system is needed. Time-division strategies such as token passing are the most common way to organize and control data signals on a baseband system; this strategy is explained later in this chapter. Baseband systems are a simple, reliable method of networking data, but can become slow when there are many nodes on the system waiting their turn to transmit or receive data.

Broadband technology is analogous to a multilane highway; many vehicles can travel, each in its separate lane, in both directions at the same time. In broadband networks a wide band of frequencies is divided into channels and simultaneously transmitted over a single cable. To prevent signal interference from scrambling the data being transferred, engineers use a process known as multiplexing. Broadband technology allows more data transfer on a LAN and is faster than baseband, but it is much more complicated because of the difficulties associated with the division of frequencies.

By combining time-division and frequency-division multiplexing techniques, we can construct very large network systems to accommodate hundreds of nodes on a single network. As the size of a DDC system grows from small to large, consideration must be given to the method of controlling data on the LAN. On very large systems a broadband network may be the more appropriate choice because of the advantage of greater speed.

Network Topologies

Networks can be organized in two ways, based on the method in which information is processed and transmitted between the nodes on the system. They are known as centralized networks and distributed networks. *Centralized networks* are based on one primary computer providing centralized processing to support multiple remote terminals. No processing occurs within individual terminals, only the input and output of information to and from the central computer.

Point-to-point networks are the simplest and earliest form of centralized network. This configuration is a direct link between the central computer and a remote terminal. *Multipoint* networks are an extension of the point-to-point concept with a host supporting multiple nonintelligent terminals on a single LAN.

A *star network* is similar to the multipoint except that instead of a single LAN cable there are many individual LAN cables accessing individual terminals.

Distributed networks provide for communication between two or more concurrently operating intelligent computers. Information traveling over the LAN can be accessed by any node on the LAN for use in processing control routines. Three common distributed network topologies are ring networks, bus networks, and hierarchical networks.

A *ring network* is a closed-loop communication LAN with nodes linked in series. The primary benefit of ring networks is that they travel very fast, and information does not "collide" with information coming into or out of the LAN. This is because each node is given the opportunity to access LAN information in turn. This orderly taking of turns is known as a token passing ring, discussed further in the section, "Network Traffic Control."

Bus networks are very similar to a party line telephone. They employ a bidirectional cable that transmits information in an out-and-back fashion. Nodes are allowed to tap off the bus at any point and can constantly monitor communication on the bus, always "listening" for information that is addressed to its specific node address. Common media used for bus networks include coaxial cable, fiber-optic cable, and broadband (also known as cable television band) systems. These media are explained later.

Hierarchical networks comprise tiers of computers that feed information to each other in a vertical configuration. They achieve fully distributed processing in that all nodes on the LAN are intelligent and produce information that is used by computers above and below them on the LAN hierarchy. No centralized processing of any kind takes place.

Figure 3-7 summarizes the topologies of both centralized and distributed networks. As you may have surmised, most LAN configurations used in modern control systems are of the distributed type because stand-alone controllers possess the power to perform local control without the need for a central processor.

There are many reasons for using highly distributed control systems for HVAC control, among which are:

- *Cost.* A LAN consists of a pair of light-gauge, low-voltage wires, which are inexpensive to purchase and install. By looping LAN wire around a building, nodes can be added to the system at any time without expensive wiring modifications. Hard-wiring between control panels is eliminated. Furthermore, inexpensive nonintelligent ("dumb") computer terminals can be connected to the LAN to access the system, thus eliminating the need to have access to a local control unit to obtain information.

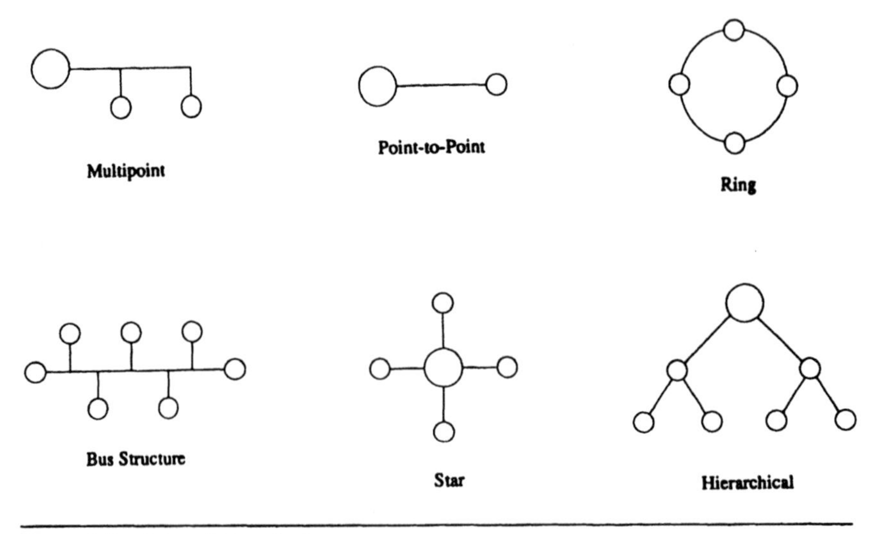

Figure 3-7 Network topologies.

- *Information sharing.* Instantaneous, or real-time, information is available to any node on the LAN for use in performing control routines. This eliminates the need for redundant field devices resulting from the remote location of a DDC panel.
- *Expansion.* The modular configuration of DDC systems allows additional points to be added to the system by adding an I/O termination board, an I/O processing card, and minor reprogramming of the existing system.
- *Maintenance.* Software is relatively maintenance free. The overall cost to maintain a computer is much lower than the cost to maintain a conventional control system providing an equivalent amount of control capacity. Likewise, a LAN requires virtually no maintenance because it is a permanently installed cable or telecommunication connection.

Network Connectivity

LANs are based on the concept of connectivity, which means that any device connected to the LAN can be addressed as an individual location. The physical connection of nodes to the LAN is accomplished by special-purpose hardware called interface cards, even though there are many names used by manufacturers to describe this device. Most DDC systems group network communication firmware and interface hardware onto a single circuit board. Interface cards provide the server with an address to refer to on the LAN when information is being shared.

Using this method, very large systems can be "localized" as a series of addresses on a LAN. By connecting individual stand-alone controllers to a LAN, DDC control can be distributed to virtually any level in an HVAC system, from the central plant right down to the occupied zone (see Figure 3-8). The only limit to the depth of this information sharing is the ability of the LAN to handle the traffic, or quantity of communication, that must move over the LAN. LANs can also be grouped together to create global networks. An example of global networking is the concept of the "smart building," where HVAC control, life safety, and security systems, each with its own LAN, are joined on a global LAN to share information through a central host computer.

A gateway is a hardware device with a software utility that allows communication to take place between nodes on dissimilar network systems. A bridge is similar to a gateway, except that it links networks that use the same technology.

Repeaters are devices that can be added to a LAN to amplify the electric signal that travels on the LAN so that longer LAN cable lengths can be used without loss of integrity of the LAN signal.

Many applications require a remote computer to communicate with

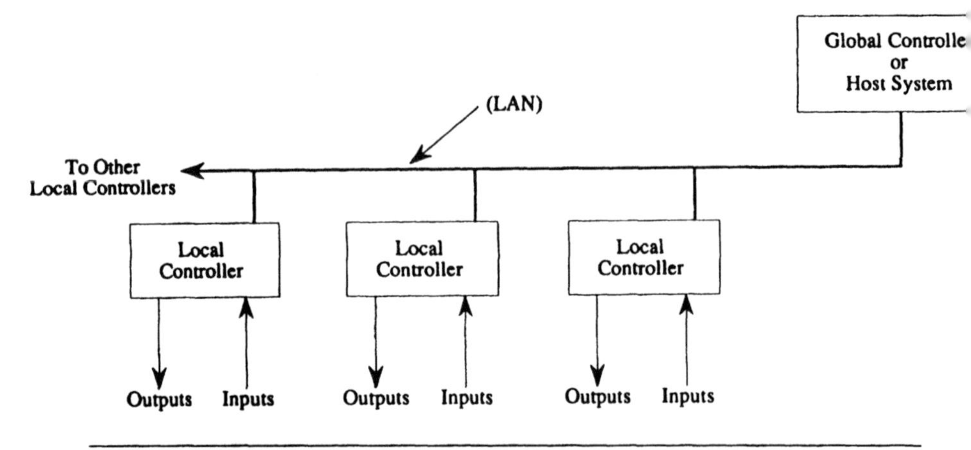

Figure 3-8 Local area network (LAN) communication.

a computer connected to a LAN. For instance, a building engineer using a PC to keep track of energy consumption in a building may require operating information from the HVAC system as an input to an electronic spreadsheet. A simple way to access this information is to connect the PC to the LAN. The communications server that controls the traffic on the network will help the PC "talk" to other computers on the LAN through a method known as terminal emulation. A terminal emulator is a software program in the server that allows a remote PC on the LAN to act as a terminal of a main computer system for the purpose of accessing information. Terminal emulation provides a communication bridge between computers that operate on dissimilar protocols. To date, terminal emulation is the only means for computer systems using dissimilar operating systems to communicate.

Network Traffic Control

A means of organizing data transmissions on the LAN is required to ensure a consistent and error-free flow of information. Because LAN systems offer the capability to support dozens of devices simultaneously, a means of controlling LAN communication is needed to prevent data from colliding. The most common collision-avoidance strategy used in baseband systems is known as token passing. Token passing is a scheme wherein a token signal is passed to each node on the LAN, giving it "permission" to transmit. The node that has the token controls the network and is the only node allowed to send a message on the network while it is holding the token. After a brief time, that node must surrender the token to the next node on the network; in this manner, nodes take turns using the network. The passing of the token and the flow of messages to and from the LAN happen very rapidly. Nodes on the LAN that have messages to transmit temporarily put them on hold while they wait for the passing of the token.

In addition to the system hardware, special network software is required to provide services for the nodes on the network. Message formatting, special message handling, and collision-avoidance routines are accomplished in firmware by burning program instructions onto ROM chips. When a device is referred to as a communication interface or communication board, it is most likely a ROM chip that is dedicated to the purpose of controlling the transmission of digital signals onto a network.

Network Media

Binary information is transmitted over network media by modulating a continuously variable alternating current signal or carrier wave. This is known as digital transmission because the voltage is coupled directly to the transmission medium. A primary concern in the selection of an appropriate network medium concerns the length of the cable and the capacitance of the wire as the distance of the cable increases. Cable capacitance has a detrimental effect on alternating DC waveforms and tends to jumble the data bits together; therefore, digital transmission is usually limited to several thousand feet of cable. The proper selection of the transmission medium is determined by the frequency and bandwidth of the signal to be transmitted. Several kinds of network cable are used in LAN systems. The three most common are twisted-pair, coaxial, and fiber-optic. The decision of which cable to use depends on the speed and reliability required for the application.

Twisted-pair wiring is simply two wires wound together to reduce electromagnetic induction and sheathed in an insulation jacket. It is less expensive than other forms of cable and is appropriate for most HVAC noncritical applications. It is susceptible to electrical interference and must be kept away from higher voltage wiring, which can cause magnetic interference. This is why power wiring and control signal wiring must always be installed in separate electrical conduits. Twisted-pair wiring cannot pass the higher frequencies associated with high-speed transmission and, therefore, is more appropriate for network systems with a limited number of nodes.

Coaxial cable consists of a central wire surrounded by insulation with a metallic braid and an outer protective covering. Coaxial cable is frequently used in broadband networks, but is also used in baseband systems. It is efficient in the transmission of signals and offers a moderate cost.

The newest medium for network interconnection is *fiber-optic cable,* which operates by using coherent light. The cable contains a small wire made of flexible glass material with devices at each end that convert the signals to a special kind of light. Signals are carried over the fiber-optic wire by the reflection of light from one point to another. Fiber-optic is at present the most expensive medium, but it is suitable for large network systems that require very high speed communication.

Network Standards

A set of standards has been developed for data communications systems that precisely prescribes the electrical and mechanical interface characteristics of the system hardware and the software responsible for controlling the flow of information between the nodes on the system. The purpose of these standards is to provide equipment manufacturers with a common set of specifications from which to design and manufacture computer-based systems.

The most widely recognized standard in data communications was developed by the Electronic Industries Association (EIA) and is known as the recommended standards or RS series. The RS series defines the way voltages are used to interconnect computer equipment with data communications devices such as modems. RS-232 is one of the first interface standards and is still in use. The term "RS-232 compatible" has come to imply that standard data communications equipment can be used with any device carrying this designation. Other standards, such as RS-432 and RS-485, were produced to overcome limitations of the early RS-232 and are concerned mostly with improvements in communication speed and accuracy.

A protocol is a formal set of rules that govern the format and timing of messages exchanged in a communications network. The *McGraw-Hill Dictionary of Scientific and Technical Terms* defines a protocol as "a set of hardware and software interfaces in a terminal or computer which allows it to transmit over a communications network, and which collectively forms a communications language."

Although transparent to the operator of the system, there are several distinct protocols working simultaneously within a computer when it is operating. The types of protocols used on a network system determine the compatibility of the network with other network systems. For this reason, most digital control systems cannot communicate with systems produced by other manufacturers. This noncompatibility is in certain cases a deliberate strategy used by controls manufacturers to prevent competing systems from being added to existing system installations, thus insuring future sales of additions and upgrades.

Seven levels of protocols have been defined by the International Standards Organization (ISO) in a model known as the open systems interconnection (OSI). This hierarchy of network functions is arranged vertically so that each level or layer can communicate with adjacent layers while operating on a network system. By separating protocols into these levels, the ISO hopes to overcome compatibility problems between systems of different manufacture by encouraging adherence to a common design standard. The seven levels are:

Level 1: physical link
Level 2: data link
Level 3: transport
Level 4: translations
Level 5: sessions
Level 6: presentations
Level 7: applications

These protocols are layered vertically on a system so that their operation is simultaneous and completely transparent to the user of the system. The seven protocols are separated into low and high levels with respect to the services each provides, and each level serves a distinct function.

The first four levels control the transfer of data and are required for any network to operate; they are usually referred to as low-level protocols. Level 1, the physical-link level, is the most important level and determines how nodes on the system physically connect to the network. Level 2 defines the way data are handled and controls the speed of transmission. Level 3 ensures that all messages are sent to and received by the correct addresses on the network. Level 4 provides translation services between networks of different types.

Levels 5 through 7 are referred to as high-level protocols or user-level protocols and are concerned with the way that the data transported by the lower levels are used. Level 5 defines the dialogue between communicating devices. Level 6 formats data for presentation to devices that have dissimilar language characteristics. ASCII code is an example of presentation-level formatting. Level 7 provides assistance to the computer operator by performing useful tasks.

Elements of a Direct Digital Controller

A direct digital controller is essentially a microcomputer, and its architecture resembles that of traditional microcomputers in almost every way. At the heart of the direct digital controller is a microprocessor that has been adapted for the special purpose of programmed control of control loops.

Direct Digital Control System Hardware

Enclosure

The power supply uses standard alternating current and then converts it into low-voltage direct current that the computer needs to drive the many semiconductor devices inside. A transformer is used for this purpose. Because incoming power is not always a level 120 VAC, protection against occasional power "spikes" is needed to prevent damage to the computer. Metal oxide varistors (MOVs) resist transient voltages in excess of their stated rating, in this case 120 VAC.

CPU Board

The *microprocessor unit* ("MPU") is the heart of the CPU. The most important component of the system is the semiconductor chip. The chips used in direct digital controllers are selected based on price, availability, processing power, power consumption of the chip, and the ability of the chip to interface with other devices in the CPU. The most common chip used in DDC machines is the Intel 8088, which, incidentally, was the chip used by IBM in its first PC.

Input/Output Termination Board

A physical point of connection between the DDC system and the sensors and control devices of the system must be provided. *I/O termination boards* or termination strips are paired screw terminals vertically arranged to allow the termination of field wiring and to provide a point of connection for the internal communication bus of the computer to sensors and control devices.

Input/Output Cards

A means of conditioning the incoming and outgoing information to the digital system must be provided so that it is understood by the digital controller. Four kinds of data can be input to a direct digital control system; the data are segregated as analog and digital through the use of I/O cards. An *I/O card* is a PC board containing electronic circuitry that translates incoming signals from the termination strip into meaningful information that can be understood by the direct digital controller. The process of converting field information into a machine-readable format is known as *signal conditioning.* I/O cards usually perform signal conditioning functions and have memory capacity to store conditioned data while the digital controller is performing other functions. This kind of memory is referred to as buffer memory.

Analog Inputs

Analog inputs are proportional or variable input signals. Examples of analog inputs are 0- to 12-VDC voltage signals, 4- to 20-mA current signals, 0- to 1000-ohm variable resistance temperature device signals, and 35- to 100-ohm variable resistance slide-wire signals.

Analog Outputs

Analog outputs are proportional output signals that are sent from the direct digital controller to controlled devices to modulate their operation. Output signals are usually of the variable voltage or current type. Other types of output signals, such as pneumatic pressure or electric pulse signals, can be generated by using signal transducers.

Digital Inputs

Digital inputs are normally contact closures where an open-circuit voltage is passed through a switch. When the switch closes and the circuit is made, a voltage will pass through the I/O termination board to indicate to the direct digital controller that a contact has been closed. Digital inputs are voltage rated because only low-voltage signals can be applied to a DDC termination board. In applications where voltages greater than 120 VAC are required for contact closures, pilot relays are used as an interface between the control voltage and the voltage of the DDC termination port. Digital inputs can also include pulse inputs. Pulse inputs are simply contact closures that are fired at a certain rate per minute or second, based on the output of a pulse-generating device, such as a kilowatt consumption meter for electric power demand measurement.

Digital Outputs

Digital outputs are contact closures that are sent from the DDC system to two-position field devices through the I/O termination board. Again, because the I/O termination board is rated only for low voltages, pilot relays must be used when loads greater than 120 VAC are being switched.

Universal Points

Whereas some digital controllers use individual I/O cards for each point type, others utilize the *universal point* concept, which means that any combination of inputs and outputs can be brought into and out of the DDC system without the need for separate cards or fixed groupings of inputs. All inputs and outputs are terminated on a single termination board, and the CPU in the digital controller may access each physical point through its address bus. Universal point systems offer an advantage over systems based on fixed-capacity I/O cards. Fixed-capacity I/O cards impose a limitation when additional input or output points are required above the maximum point capability of a card. For instance, an analog input card that can accommodate eight inputs cannot accept a ninth input without the addition of an additional analog input card. In DDC systems using a fixed-card architecture, the capability of accepting an extra card to accommodate additional points may not be available, forcing the user of the system to purchase an additional DDC control panel to add a small number of points. This has proved to be an economic disadvantage for owners of fixed-architecture systems in the past.

Transducers

Transducers are devices used to convert control signals from one form to another. When DDC systems are integrated with pneumatic control devices, input and output signals must be converted from pneumatic to electronic and electronic to pneumatic, respectively. Many DDC panels are provided with input and output transducers as a built-in feature. Others require outboard transducers on incoming and outgoing signals. Once pneumatic control signals have been converted to analog format, they must then be converted from analog to digital format to be processed by the direct digital control or microprocessor. Analog-to-digital (A to D) converters perform this task so that digital information can be evaluated and transmitted to the main CPU. Digital transmissions from the CPU are converted to analog signals, which can then be used to control analog field devices or they can be converted back to a pneumatic signal for the same purpose.

Communication Ports

For individual DDC panels to communicate over LAN or to interface with other types of peripherals, such as printers or video terminals, connection ports are required. The most common type of connector is the RS-232 serial port. All interfaces to peripheral devices are made through this port. For the CPU to recognize the peripheral device, its address must be registered. This is done by arranging dual in-line package (DIP) switches in a proper configuration. Some systems opt to use a selector switch in lieu of DIP switches, but the purpose is the same. It tells the CPU where to locate the peripheral device on its communication bus.

Usually ports for peripheral devices and for connection to the LAN are kept separate. LAN communication ports are usually of the RS-485 type, which provides for very high speed communication to take place over a twisted pair of shielded wires. Each device, or node, on the LAN must have a specific address so that other nodes on the LAN can recognize each other and allow the LAN server to recognize each node on the system. This is accomplished by utilizing DIP switch configurations. Each node has a different DIP switch configuration, which is simply a series of on and off positions across an array of eight switches. Communication address switching may differ between manufacturers, but the purpose is the same—to provide a specific value so that data can be transferred between each node on the system and/or a host computer.

LAN Protection

LAN wiring is susceptible to electrical interference from nearby electric loads, such as lighting or power distribution, or from transient surges of excessive voltage caused by lightning strikes. To prevent serious damage to the network wiring circuit and the devices connected to it, a means of circuit protection is needed. A transient surge arrestor, an electrically grounded device designed for low-current signal loops, is installed between the LAN and each node connected to it. These arrestors are relatively inexpensive and highly recommended for installations in areas where the potential for damaging signal interference is possible.

Firmware

Firmware describes software programs that permanently reside on microchips in the DDC system. Intelligent DDC panels normally contain several firmware chip sets, each serving a different purpose. Permanent software such as the operating system, utility programs, and some application programs reside in firmware.

Operating System

The programs that tell the digital controller how to operate are called the operating system. These programs usually reside in ROM chips and are specific machine-level instructions that only the microprocessor chip can understand. The operating system operates transparent to the user.

User Interface

User interface describes the many devices that allow humans to access the information in the control system. Common interface devices are keypads, monitors, panel-mounted liquid crystal display (LCD) readouts of information, and hand-held operator terminals. Interface can occur at the local field panel level or at the central level in a hierarchical system through a host computer.

LCDs are common at the field panel access level and provide limited information to the operator through special function keypads. At the host level, however, active screens on a computer monitor can provide current values for control points throughout the system. The host system collects data from each field control panel as it polls the system for current operating data.

A detailed discussion of user interface is provided later in this chapter.

Direct Digital Control System Software

There are two kinds of programs used on microprocessor-based control systems: system programs and application programs. *System programs* are responsible for helping the microprocessor form its tasks in an organized manner. *Application programs* are defined-purpose programs that tell the computer how to perform the work it is supposed to accomplish. Although these two types of software are interrelated, they are distinct. System programs, also called operating systems, are limited to telling the computer how to operate within its own environment. The operating system tells the microprocessor how to interpret data, how to communicate with other devices in the direct digital controller, and how to interface with the outside world. Application programs, on the other hand, tell the computer how to perform control routines that provide a meaningful output to a human operator. Operating system software is permanently stored in firmware in DDC panels and often in application software programs as well. This is because direct digital controllers are frequently exposed to power failures and other forms of electrical interference. If these programs were not permanently stored in ROM, they might accidentally be destroyed by a power failure or operator error.

Utility Software

Utility programs perform functions that support the operation of the microprocessor and the application software that is running on it, and are normally built into the operating system directly. An example of a utility program is a routine that performs a diagnostic check to make sure that all of the devices under the control of the operating system are performing correctly. Another example of a utility routine is the automatic copying or backing-up of the DDC system database to an external storage medium such as a floppy disk or a hard disk drive. The need for utility functions largely depends on the size of the control system and its architecture. A system comprising many subsystems generally requires more utility routines than does a simple microcomputer. Most DDC systems have a self-test function that is executed when the computer is powered up initially. Self-test functions normally check CPU operation, confirm the integrity of communication between the CPU and its attached I/O devices, make sure that there is enough RAM memory available for proper operation of the computer, and double-check the integrity of the database. The execution of utility routines is invisible to the operator and normally takes place very quickly after the system has been initialized.

Application Software

As stated earlier, application programs carry out the specific tasks that we want done. With respect to DDC systems, application programs execute specific energy conservation and supervisory control routines to improve the performance of an HVAC system. There are virtually no limitations to what can be done in application software so long as the system hardware is able to support the routines being performed. That is, as improvements are made in the way HVAC systems are controlled, new application programs can be written and applied to DDC systems with very little impact on the hardware that supports the software. There are no universally applicable control routines that will satisfy every building condition, but there are some proven strategies that, when modified to meet the needs of a building, are effective in reducing energy consumption. Several of the more common software routines used in modern building control systems are described next.

Energy Management Software

One of the most basic and successful energy management strategies is to schedule the operation of electric motors in HVAC systems. By scheduling equipment to run only when it is absolutely necessary, kilowatt consumption can be reduced significantly. Also, by staggering the start times of large motors, heavy in-rush currents can be staggered to reduce the overall peak load of an electric system.

EMS software is used to control and monitor each piece of equipment in a system, including its start time, stop time, and current operation status. By connecting a printer to the DDC system, we can keep a log of all start/stop times to ensure that the system is performing as required. Application-specific energy management software is described in greater detail in Chapter 6.

Man–Machine Interface

The apparatus that brings data from within the control system to its human operator is referred to as the *man–machine interface* (MMI). Through the MMI, a building operator can interrogate the control system for information on how the system is operating, and he or she can initiate commands that alter the way the control programs are performed. MMI can be located at the host level or at the local field panel level, depending on the capabilities of a given system. It is quite common to find an MMI at every intelligent control panel in one form or another; keypads with LCD readouts are quite compact and easily fit within the space constraints of a digital control panel door. Most systems provide only a limited amount of information at the local level, which is accessed by pressing special function keys that bring information to the display screen. As control technology has developed, the quantity and quality of local information have increased dramatically. Some digital systems communicating over high-speed LANs can now bring information from every point in the control system to any MMI point on the LAN.

Information to an operator can be displayed in textual or graphic formats as computer code strings, menu-driven English language prompts, or easy-to-understand graphic screens.

Encoded information is machine-level information and is of use only to the system programmer or analyst. Codes are usually in hexadecimal format or in a machine-readable format that is proprietary to the control system. Menus provide user information through a fill-in-the-blank format that is usually in English or a similarly abbreviated form. When the desired information from a preprogrammed menu is selected, the machine can retrieve preformatted information for the operator. This feature is useful for monitoring ongoing operations and for diagnostic checks in the event of a problem.

Graphic information is the most user-friendly form of information. Preformatted screens depicting equipment and control system arrangements provide the operator a visual display of system operation. Active fields of information on the screen are updated (also known as refreshed or populated) by the host computer, which is continually polling field devices for current values. Graphic screens and their active fields are established when the system is initially programmed.

Information from the system is accessed by point attribute. This is how the computer database defines the nature of control points and how the computer searches for the information it is being asked to gather. Point names (electronic addresses within the computer system), point descriptions (the location of points on the LAN [node locations]), normal operating conditions for each point, and point alarm limits are examples of point attributes. Point data are entered into the database and organized through system programming, which is discussed in the next section. Several important user functions are performed through the MMI. They are prioritized by user access level to prevent accidental or deliberate tampering to the database. Most intelligent control systems provide several levels of user access that are segregated by access codes or passwords. The issuance of passwords can be controlled by the building management to prevent unauthorized access to the control system, be it local access by operations personnel or through telephone access to the system by modems. The first four functions are considered low-level access functions, whereas the others are usually restricted to those who are trained in the programming and maintenance of the control system database.

The eight most common MMI functions, ranked by priority level, are:

1. Monitoring point data
2. Acknowledging alarms
3. Temporarily overriding commands
4. Creating reports
5. Creating alarms
6. Creating new programs
7. Modifying existing programs
8. Maintaining the database

Accessing current point values is the most common MMI function. Points on the control system can be called up on the screen for visual verification. In some systems, groups of related points can be simultaneously reviewed to assist the building operator during troubleshooting.

Alarm acknowledgments tell the control system that a reported alarm has been recognized and handled by building personnel. Acknowledgments are usually handled by answering a screen-prompted message through a few keystrokes, and they are usually required to silence an audible alarm initiated by the control system. Most systems provide a printed log to serve as a record of when an alarm occurred, where it occurred, the current values of the alarm point when reported, and the date and time that the alarm was acknowledged by an authorized operator.

Temporary control overrides are used to allow building operators to change temporarily certain parameters of a control program. Time schedules and set points are commonly overridden to accommodate special events. Override command time limits are usually programmed into the database.

Creating reports is one of the most important features that a digital system can offer to a building operator. By selecting specific points to be monitored over a given time frame, an operator can observe the behavior of the control system to look for trends or solve a control problem. This is usually a low-level access function because it is a tool for improving system performance through analysis; its access should not be restricted from operators who are responsible for building operations and maintenance.

Creating alarms is done by adding limits to the point attribute in the database. For example, the desired set point for a room being monitored by a thermistor sensor may be 70°F, with an upper limit of 78° and a lower limit of 68°. When the sensed value exceeds these limits, an alarm is generated. This access level also allows alarm limits to be modified and should therefore be restricted to experienced system operators only.

New programs are created when new points are added to the system or when a new control strategy is to be added to the application programs. New graphic screens are created at this level, and the active point value fields are entered into the graphic. Point attributes are established and specific operating instructions are given to the database to be stored in permanent memory. Existing programming and graphics can also be accessed at this level when changes in operating routines or graphic displays are needed.

Periodic maintenance of the database should remain at the highest level of system access. Database maintenance entails adding new points to the system as well as adjusting the attributes of existing points to keep the database current. This should be done periodically, and a hard copy of the previous and current programming should be kept to record significant changes to the software. A backup of program codes should be kept for all levels of software, including the operating system and application programs. This is usually kept on floppy disks or magnetic tape for easy reinstallation in the event of a system crash.

Direct Digital Control System Programming

Programming and data entry to the control system database are usually done at the host level through user devices such as the keypad, mouse, light pen, or touch-screen. The available user devices differ among control system manufacturers, but all serve the same basic purpose—to allow a simple and friendly way to instruct the system how to operate.

In highly distributed systems, programming can be done at the local control panel level and can be uploaded or downloaded, as the case may be, throughout the network of controllers to any point of intelligence in the control system. Typically, major programming and data entry are done at the host level and are downloaded through the system, whereas minor control program adjustments and point updates are done at the local level and uploaded to the host programs.

The degree of complexity encountered in initiating programming is related to the computer language used to interface human programmers to the system. Some programming languages are simple, and others are quite difficult for a nonprogrammer to understand. In the past, programming complexity usually followed system capability; the more difficult the language, the more flexible and powerful the programs it could create. This is no longer the case. User-friendly program languages now available are layered over the computer machine language to translate control sequences into algorithmic form.

Three major types of programming languages are used for digital control systems: line-oriented programming, custom line-oriented programming, and control block-method programming.

Line-oriented programming (LOP) uses a high-level computer program such as BASIC, FORTRAN, or Pascal to create the control sequences. The program code is defined in a line-by-line format as shown in Table 3-3. Each control sequence is written in a separate subroutine. Anyone familiar with the programming language can understand the control programming after a brief orientation. Some line-oriented languages have been modified to utilize a fill-in-the-blank format, which reduces error on the part of the computer programmer and also makes it easier for programmers who are not familiar with a particular programming language.

Custom-line-oriented programming (CLOP) is similar to line-oriented programming; however, the formats used in the programming language are proprietary. Although the purpose of custom line-oriented programming is to make the practice of programming easier for control technicians and building operators, it requires a thorough knowledge of the structure and mnemonics of the language. Table 3-4 illustrates a common CLOP program.

The third type of programming method is known as the *control block method.* With this method, little or no line-by-line programming is used.

Control sequences are individually specified in logically organized "blocks" that define the nature of the input signal, the parameters of the control function to be performed, and the nature of the output signal. Parameters for each block are entered using the fill-in-the-blank format shown in Figure 3-9. Each control loop is given a unique name and information is entered that describes exactly how each control loop should perform. After control blocks for each control sequence

Table 3-3 Line-Oriented Programming (Basic)

Line #	Code Text
10	CLS:PRINT "What is the message to display?":PRINT
20	LINE INPUT MSG$
30	PRINT: PRINT "Type 'Q' to end the screen saver. . ."
40	FOR J = 1 to 25000
50	J = J+1
60	S = LEN (MSG$)
70	ROWNUM = 4 : COLNUM = 7
80	CLS
90	LOCATE ROWNUM, COLNUM
100	Print MGS$
110	L = m4.1
120	ROWNUM = INT(RND*20)+1
130	COLNUM = INT(RND*(80-S))+1
140	WHILE L
150	PRINT TIMES
160	L = L + 1
170	IF L = 190 THEN L = 0
180	A$ = INKEY$
190	IF A$ = 'q' OR A$ = "Q" THEN 220
200	WEND
210	GOTO 80
220	CLS
230	SYSTEM

Table 3-4 Custom Line-Oriented Programming

```
Routine #:           30
Message/Title:       ROUTINE TO ACTIVATE AHU CONTROLS
Active (Y/N):        N
Preactivate (Y/N):   N
A      If [AH PROOF; CV]=1
B      Activate routine #28
C      Activate routine #22
D      Make [AHU MA]=3
E      Make [AHU SA]=3
F      Deactivate routine #34
G      Make [RETURN; AHU]=760
H      Make [ZONE; AHU]=760
I      Make [ECON; AHU]=0
J      Make [COIL; AHU]=0
```

Source: Barber-Colman Company

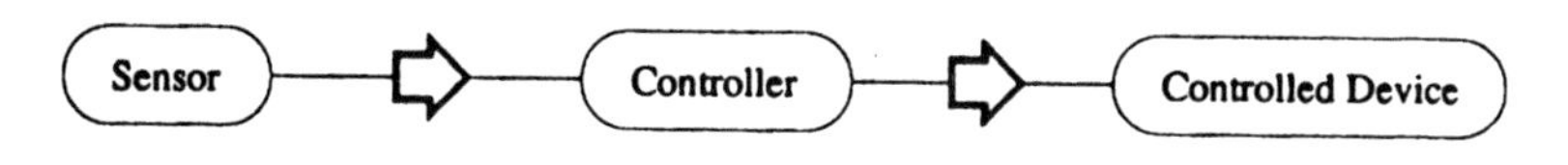

Figure 3-9 Control block layouts versus pneumatic control diagrams.

in a system have been defined, the blocks can be linked to create complex and complete control sequences. Control block layouts are very similar to the pneumatic control diagrams shown in Figure 3-9.

The purpose of block programming is to allow programmers with a strong background in HVAC systems but with little background in computers to easily and intuitively program. Furthermore, when properly documented, this form of programming provides a complete record of the control sequences being performed by the control system. Most digital control systems now use block method programming in one form or another.

Chapter 4

Interfacing Digital Controllers with Conventional Control Devices

This chapter discusses the input and output functions of a digital control system. These functions, known as field data acquisition and output communication to controlled devices, are the primary determinants of how effectively a digital control system can operate. Data acquisition is concerned with how information is gathered and sent to the digital controller; output communication involves the methods used to convert control commands from one format to another to allow devices with incompatible control signals to work together. Chapter 2 introduced control system components and briefly described sensors and controlled devices. This chapter explains how these devices are connected to the digital controller and describes the factors that affect system performance.

Signal conditioning is a process that improves the quality and accuracy of a low-intensity control signal to make it readable by a controller. This function is applied to incoming data as well as outgoing control command signals. Figure 4-1 illustrates the path that unprocessed analog and binary signals travel on their way to the central processing

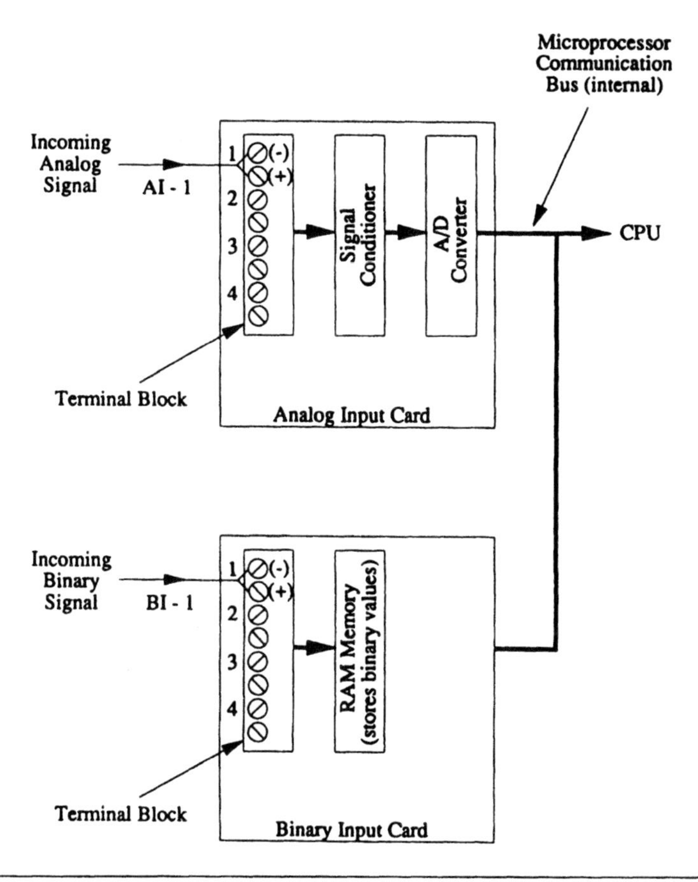

Figure 4-1 Field data acquisition.

unit (CPU). Signal conditioning is necessary because a digital controller assumes that the information it receives is error free. For example, a resistance signal that is of low intensity, such as a 1000-ohm Balco sensing element that changes its resistance at the rate of 2.2 ohms per °F, is very susceptible to electrical interference. The added resistance of the wires that carry the signal to the controller weakens the signal and skews its accuracy.

Another example of a low-intensity control signal is a static- and velocity-pressure signal from an airflow measuring station. Because these signals are so small, even the slightest agitation, electrical or mechanical, can ruin the value of the signal and the ability of the controller to operate an HVAC system. Transmitters were designed to provide the functions of signal conditioning and transmission to eliminate this problem.

A *transmitter* is a device with an electric energy source that is capable of amplifying low-intensity signals and transmitting them accurately over long distances of wire. Signal amplification reduces the effect of wire resistance to the point where such resistance can be corrected for by the digital controller. A transmitter is usually located near its sensor so that the sensor lead wires can be short to avoid signal error. There are three types of transmitters—current, voltage, and resistance—based on the type of signal they use to transmit information to the digital controller. Most transmitters are also transducers, because they convert a control signal from one format to another in the process of transmission.

Transmitters are calibrated to provide accurate and repeatable signal transmission, and they are usually adjusted to obtain the most accurate signal possible given the offset and span requirements of the controlled media.

Offset is the input value required to produce a predetermined transmitter output. Think of offset as the starting point of the output signal. Span is the total change in the input value required to produce the maximum transmitter output. To illustrate the functions of offset and span, consider a room sensor and a current transmitter as an example. The temperature range of a room over all seasons is determined to be 10°, from 68 to 78°F. A resistance temperature device (RTD) sensor measures room temperature, and this signal is transmitted back to a digital controller through a current transmitter, which converts the resistance signal to a current signal for transmission to the controller. The output signal of the current transmitter is scaled from 4 to 20 mA. Therefore, the offset (starting point) of the transmitter is 4 mA at 68°F. The span of the signal is 10°F, which means that at 78°F the transmitter output signal will be 20 mA. Calibrating adjustments to offset and span values are made either at the transmitter or from the digital control panel.

Current transmitters are manufactured with a 4- to 20-mA output range as the input signal varies from its minimum to its maximum, according to ISA Standard S50.1 (1975), which standardized the 4- to 20-mA current output signal range among manufacturers of current transmitters.

Input Functions

Two types of input signal formats can be processed by a digital control system: analog (proportional) and binary (two-position).

Analog inputs range in value from a predetermined minimum and maximum, such as 4 to 20 mA, and are calibrated to represent the range of the controlled media, such as 68 to 78°F.

Digital inputs are two-state signals, analogous to a digital switch in an open or closed position. Digital input signals are used to indicate the operational status of HVAC equipment and, when used in a pulse-width-modulated (PWM) format, can be used to create a synthetic analog input. Pulse-width modulation is a technique that uses a digital signal to create an analog signal by modulating the duration of a switching action. For example, a pulse signal that modulates open–closed–open once every 0.1 millisecond can represent a specific value to the digital controller. A common application of PWM digital inputs are pulsemeter inputs from current transformers used to indicate power consumption. As the pulse signal duration decreases, this indicates a proportional increase in the kilowatt consumption of an electrical device.

For the digital controller to be able to process the incoming signals, signal values must be converted from analog or digital status into binary numeric values. This involves converting values from decimal (base 10) to binary (base 2) notation. Data conversion normally occurs within an interface device that resides between the incoming control signal and the CPU of the digital controller.

Interface Devices

Digital control systems are usually equipped with interface cards (commonly referred to as point cards) that convert incoming control signals and condition them before they are sent to the CPU for processing. Interface cards are usually designed in a proprietary fashion to serve only the system they were manufactured for; usually these cards are not interchangeable with other digital control machines.

Point cards are usually configured by point type. Analog input, analog output, binary input, and binary output cards are mounted in an electronic card rack in the digital controller enclosure, and the points are usually distributed in even increments. For example, a 32-point digital control panel will usually house four cards of eight points each. To add flexibility to the point limitations of a controller, many manufacturers offer input cards that support more than one type of point. For example, a single eight-input card may have four analog inputs and four binary inputs; conversely, cards can be dedicated to only one type of point, such as eight binary inputs. Properly designed input cards offer design engineers and controls contractors the flexibility to accommodate most control system point configurations efficiently.

Often there are situations where the quantity of the input data exceeds the input capacity of the digital controller. This would require the added expense of additional control panels and associated installation and wiring costs when the input point density exceeds the point limits of a system or control panel; in this case a device known as a multiplexer can be used. An input multiplexer enhances the flexibility of a digital controller by allowing the controller to sequentially access multiple input points, each with a separate address. For example, say we are multiplexing 10 separate zone temperatures through a single analog input channel by switching from one signal to another. The application program running on the digital controller can access information on the temperature in, for example, sample zone 5, through an address on the multiplexer. Even though there is only one analog input value, it can select which value to read through a digital switching network.

Data Conversion

Information from the input cards is converted to digital information by an analog-to-digital converter in the digital control enclosure. The input signal is first converted from a voltage or current value into a decimal number. This is usually done by the logic circuits on the input card. The decimal value is next converted into binary notation, which can then be interpreted by the CPU. This is where the true power of digital control systems becomes evident. Because 4 bits equals 16 numbers (4 × 4), a single 8-bit byte equals 256 numbers (0 through 255). Such a wide number range is advantageous when an input control signal is scaled because the increments of change can be very small.

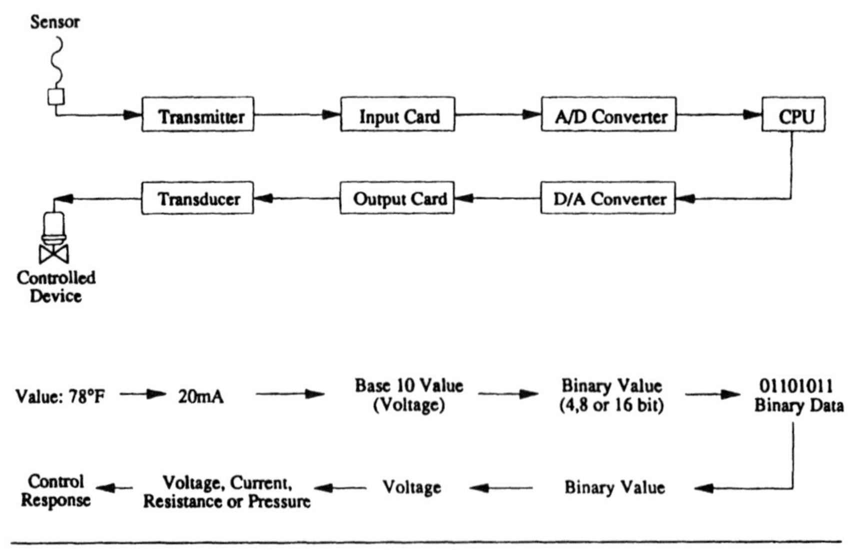

Figure 4-2 Input/output data-conversion path.

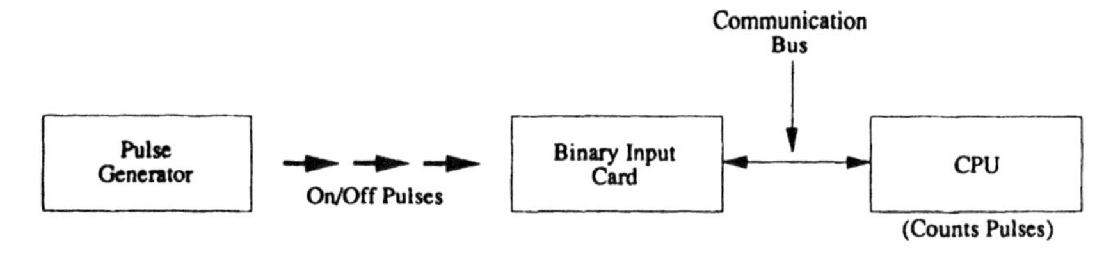

Figure 4-3 Binary-pulse data-conversion path.

For example, the difference in the input signal value caused by changing the converted value by 1 bit produces a 0.4% change in the range of the sensor (1 divided by 256). This increment is known as the *precision of the conversion.* This example assumes that the digital control machine utilizes an 8-bit format. There are digital machines available with 16- and 32-bit processors that offer even greater precision: the larger the word length, the greater the precision. Figure 4-2 illustrates the path that an input signal follows from data conversion to the CPU.

Binary inputs have a data conversion path similar to binary format. A binary input card routes the input signal from the termination board through the input card and generates a signal to the CPU in byte format. Usually this data conversion takes place within the binary input card because there are only two possible control states. Binary input cards usually contain RAM chips to store binary values for the CPU, which can then access these data when needed. This memory chip is also referred to as buffer memory, because it stores information for future retrieval. In addition to change-of-state signals, binary inputs can be in the form of intermittent pulse-width modulations. By counting the quantity and time duration of pulses, we can derive a synthetic analog value. This value can then be converted from this quasi-analog state into virtually any other type of value the digital controller needs. Figure 4-3 depicts the binary-pulse data-conversion paths through a digital controller.

Output Functions

The output signals from digital controllers can be analog or binary. Analog outputs of variable-voltage, variable-current, and variable-resistance type are common. Binary outputs can be on/off contact closures or pulse-width modulations.

Analog Outputs

When output control signals are generated by the CPU and the digital controller, they must be converted from binary format to their original format through a digital-to-analog converter. In this process, bytes of information are converted to voltage or current values and are routed by an interface card to the output channel that corresponds to the address of the device the signal is intended to control. This routing is established in application software when these programs are installed in the system. Once this newly converted signal reaches the output channel, it is then conditioned and sent to the control device.

Often the output signal that the digital controller generates is not compatible with the input signal requirement of the controlled device. Therefore, a means of transducing the output signal into a correct format is required.

Output Signal Transducers

Most digital control systems are capable of generating either voltage or current output signals, in some cases both. In situations where a digital controller is limited to one type of signal, a transducer is needed. *Voltage-to-current* (E–I) transducers convert variable voltage signals into variable current signals. *Current-to-voltage* (I–E) transducers do the reverse.

The most common transducer used with digital control systems is the analog-to-pneumatic transducer. This is used to digitally control pneumatically actuated valves and dampers. There are two primary types of analog-to-pneumatic transducers. The first type is the *current-to-pressure* (I–P) transducer. This takes a variable current signal, such as 4 to 20 mA, and converts it to a corresponding 3- to 15-psi output signal. The second type is the *voltage-to-pressure* (E–P) transducer. This device transduces a 3- to 15-VDC signal to a corresponding 3- to 15-psi output signal. In the same way that a binary input card can accept a PWM input signal, a digital controller can produce a PWM output signal through a binary output card. There are three types of output signals that can be generated with a PWM signal. They are categorized as resistive signals, multiplexer signals, and PWM-to-pneumatic conversions.

PWM output signals are normally used when only binary output signals are available to control a proportional device. The accuracy of a PWM output signal depends on the number of pulses that can be fired within a given time increment. Millisecond switching provides near perfect analog control: the more pulses per time increment, the greater the control accuracy. Figure 4-4 illustrates the conversion from PWM to voltage, which increases in conversion accuracy as the pulse duration increment decreases; increments of 0.05 milliseconds offer more accuracy than conversions made in 0.1-millisecond increments. PWM switching is accomplished within the digital controller by toggling miniature relay contacts on the binary output board. These con-

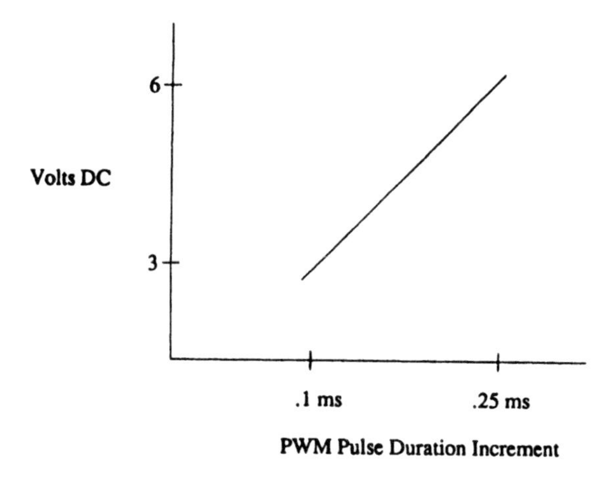

Figure 4-4 Pulse-width modulation to voltage conversion.

tacts, made of fine metals, such as gold and silver, can be switched millions of times without wearing out the contacts.

Proportional control of HVAC equipment, such as fan discharge dampers or inlet vane actuators, is performed with resistance-type slide-wire controls. Slide-wire control uses a potentiometer to vary the resistance of an electric circuit to modulate the speed of an electric device. A familiar example of a slide-wire control is a rheostat that dims a light bulb when it is turned. A PWM output from a digital controller can simulate a proportional slide-wire signal through a PWM-to-resistance transducer. This device, also referred to as a *digital resistance network* (DRN), produces a variable-output resistance signal that is proportional to the time duration of the PWM input signal. This is done by selecting the output resistance value from a network of resistors built into the DRN transducer. Through groups of resistors of different ohm values, a DRN transducer will switch through its resistor network until it matches the ratio of the PWM control signal conversion. DRN transducers are available with very fine steps of resistance to provide highly accurate output control signals from a PWM signal (see Figure 4-5). The same effect can be accomplished by DRN transducers that accept current or voltage input signals as well.

After a digital control system is installed, it is common to find that the building owner would like to add other loads to the system. Too often, the point expansion to an existing system can become quite expensive on a per-point basis because additional hardware must be added to the system to accommodate even a few additional points. A low-cost means to expand the binary output capacity of an existing system is to install a PWM multiplexer. These are especially useful

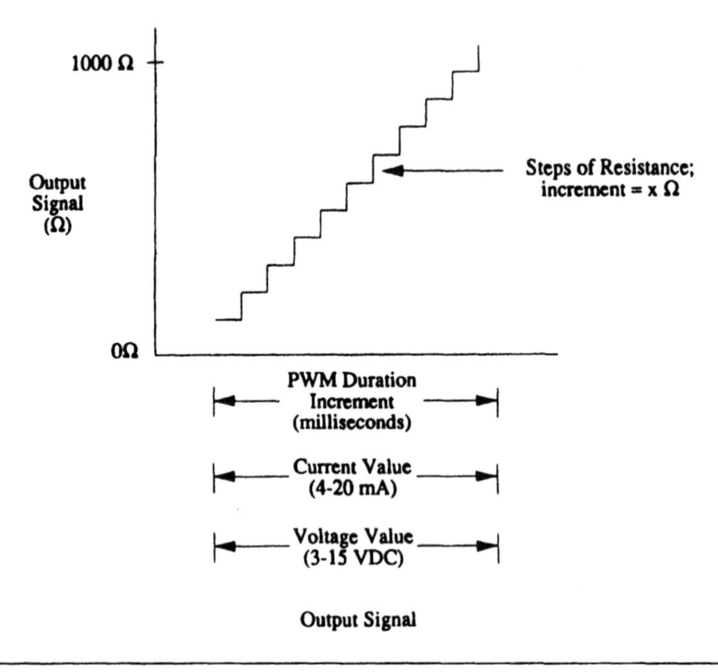

Figure 4-5 Pulse-width modulation to resistance conversion.

in applications where more than one binary device is to be controlled, such as time-scheduled switching of multiple lighting loads. An output multiplexer card expands the output capacity of the digital controller by providing several relay outputs that can be switched by a single PWM signal. Each output switch, or relay, is assigned a PWM duration increment value that will toggle the output. Figure 4-6 charts this switching sequence for an eight-output binary multiplexer.

A multiplexer can also be used to increase the analog output capability of an existing system by providing a pulsed output as an input to another digital controller. For example, one of the output channels of a multiplexer can act as an input signal to a PWM-to-voltage transducer, which can then proportionally control an electronic valve actuator. By taking advantage of the relatively wide PWM duration increment range, we can activate many switching sequences from a single digital output at the digital controller. Pneumatically controlled devices are still popular because of their ease of maintenance, high reliability, and low cost. To interface a digital system to a pneumatic device, we need a special type of output transducer to convert an analog or binary signal into a variable pressure signal. An electronic-to-pneumatic (E–P) transducer provides this interface by converting a variable voltage, variable current, or PWM signal to a corresponding output pres-

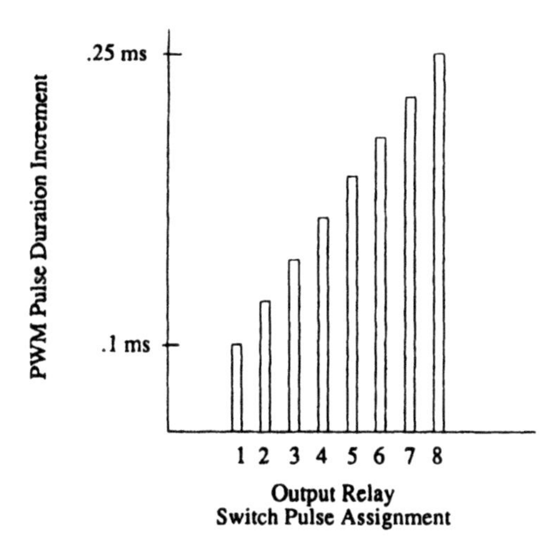

Figure 4-6 Pulse-width modulation to output multiplexer conversion.

sure signal. This device requires a source of main supply air pressure to the transducer. The output pressure to the controlled device is mechanically controlled by a proportional air valve that modulates the output pressure signal in proportion to the input signal. A pneumatic feedback signal to the transducer ensures that the proper output pressure is being provided. E–P transducers are available in direct- and reverse-acting configurations. Current-to-pneumatic (I–P) output transducers are also available. Both types of transducers are available as inboard (built into the DDC enclosure) or outboard devices. Most use a built-in microchip to self-perform signal conditioning.

Binary Outputs

Digital information from the digital controller is transmitted to two-position controlled devices through an output channel switch closure. *Field-effect transistors* (FETs) are transistor switches that reside in a binary output card and are the means by which binary signals are achieved. The voltage and current capacities of binary output relay contacts are very low and are not intended to switch other than low-load control circuits. When an application requires that the binary output signal switch a 24-V, 120-V, 277-V, or higher rated voltage circuit, an interface device is required.

Interface relays, or pilot relays, are used to bridge the binary output channel signal to a two-position controlled device. For example, when 277-V, single-phase loads are switched in a lighting control panel, a 24-V rated pilot relay is used that has contacts rated for 277-V power.

The output termination strips on some digital control products are equipped with built-in pilot relays capable of switching up to 240-VAC circuits safely. Care must be taken to make sure that a binary output does not directly switch a load in excess of its rated capacity, or else severe damage to the digital control system can occur.

Relays can be operated in a two-position mode, a time-duration switch mode similar to the PWM technique (although PWM through pilot relays is not recommended owing to the resulting fatigue on the relay contacts), or by latching circuits that hold a relay in position as long as current is applied to a magnetic coil. This is also known as momentary switching. Familiar uses of relay control are lighting circuits, security sequences for building access, and life safety sequences for alarm initiation.

Interface relays are almost always required to start and stop electric motors on fans and pumps. Electric motor starters are usually equipped with a transformer and a control voltage relay. When voltage is applied to the control relay, normally open relay contacts are closed to allow power to pass through the starter and into the motor. Figure 4-7 presents a ladder diagram of a simplified motor starter using a control relay for start/stop control. Binary output switches can also switch electric solenoid air valves to provide controlled air to pneumatic control systems. This method is frequently used to sequence pneumatic control modes. A common example is to switch from a morning warm-up mode to a normal mode of operation in combination digital/pneumatic control systems. Usually an interface relay is not required for a binary output channel to switch an E–P solenoid because the contact rating on the solenoid is usually quite low.

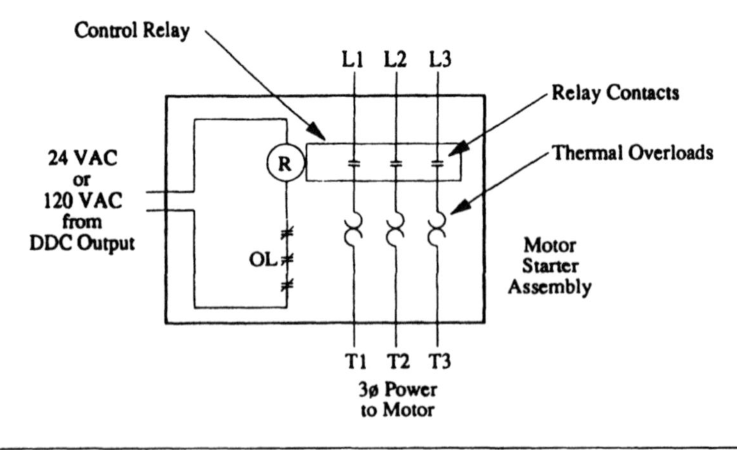

Figure 4-7 Motor starter control using control relay (simplified).

Motor speed control can also be accomplished with a binary PWM output signal from the digital controller. Most variable-frequency drives (VFDs) accept a PWM signal as an input signal to its control panel. The microprocessor within the VFD will count the PWM duration increments and will vary the rpm of the motor in accordance with the PWM pulse rate. The PWM signal must be scaled in increments that exactly match the input signal scale of the VFD input channel. Most VFD products also accept voltage or current input signals, which may also be generated from a digital output point through transducers, as discussed earlier.

Interconnecting Media

Interconnecting media refers to the wiring systems that are used to connect the components in digital control systems. There are three situations where external wiring systems are required: power wiring systems, signal wiring systems, and communication wiring systems.

Power Wiring Systems

The type and size of wire used to provide electric power to control devices is determined by the quantity of current and voltage they require. Most digital control devices require 120-V single-phase AC power or 24-V DC power. By summing the VA (volts times amps) requirements of all control devices, we can properly select the type and size of wire.

Because power wiring systems are material- and labor-intensive, field-mounted step-down transformers are commonly used to bring power to local control devices. Figure 4-8 illustrates a typical power wiring system using step-down transformers. In this example, 1000 A of 120-V AC power is distributed throughout a building. Individual 120- to 24-V step-down transformers are branched off of the main power loop to bring the correct voltage and amperage required to each device. If a single power wiring loop were used to power these devices in series, a much larger power conductor and conduit would be required. The reduced size of the main power circuit more than offsets the additional cost of furnishing and installing control voltage transformers at the local level.

Usually, the power requirements for control devices can be satisfied with relatively small gauge wire; 18- to 22-gauge is common. Local codes should be checked to determine the materials and methods required for wiring systems. Once the type and size of wire have been

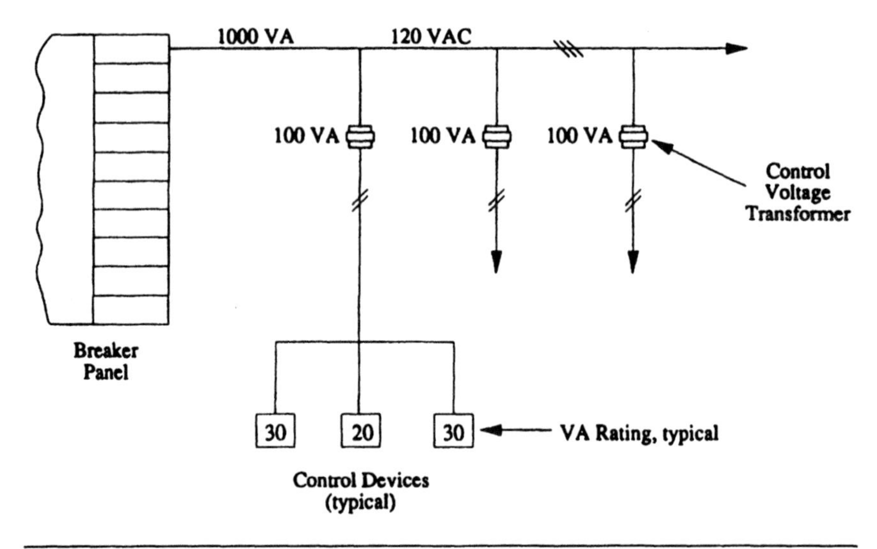

Figure 4-8 Power wiring systems.

selected, the type of raceway that must enclose the wiring must be selected based on the conditions to which the wire will be exposed. A *raceway* is an enclosed channel designed for holding wire and cable and is usually constructed of metal or insulating material. The most common raceway used for temperature control system wiring is called electrical metallic tubing (EMT). The National Electrical Code prescribes the maximum capacity of conductors that can be contained in a raceway and the minimum requirements for conductor protection.

Signal Wiring

The most critical wiring in a digital control system is the wiring that carries control signals to and from sensors and control devices to the digital controller. This is because the control system can only operate as accurately as the integrity of the incoming and outgoing control values allows it to. The resistance of electric wire must be given careful consideration when it is used to carry low-intensity signals back to a control panel because the resistance in wire increases in proportion to its length. For example, a 100-foot length of 18-gauge copper wire has a resistance of 0.64 ohms. Although small, this resistance can skew the accuracy of a voltage signal that is being transmitted over this 100-foot length of wire. For resistance-type sensors, such as RTDs and thermistors, this added wire resistance creates an error that affects the accuracy of the signal received by the digital controller. Transmitters are used to offset signal losses by amplifying the sensor signal

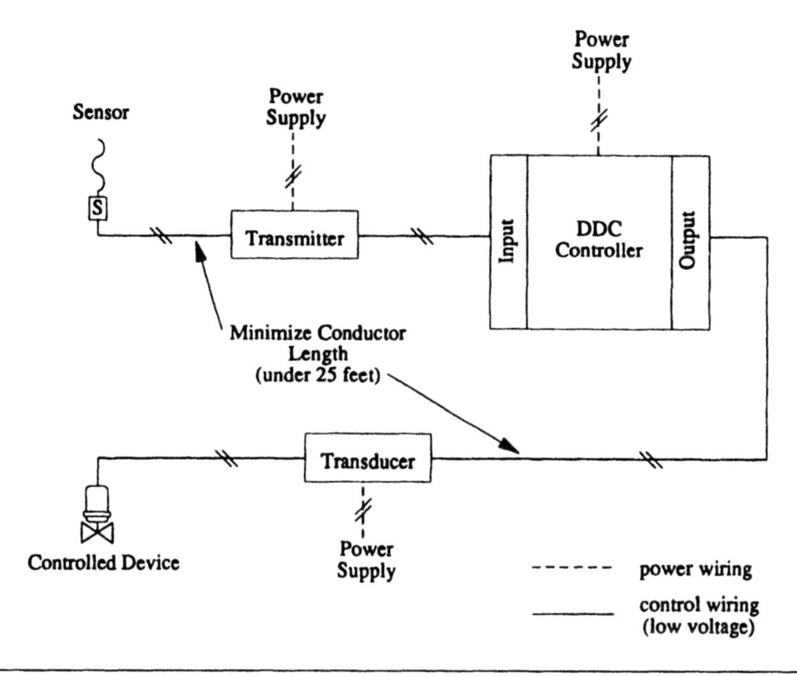

Figure 4-9 Sensor wiring systems.

and converting it into a voltage or current signal which is then sent to the controller. Therefore, the distance between the sensor and its transmitter should be kept as short as possible. Most sensors now incorporate a built-in transmitter to eliminate this distance.

Likewise, the wiring between the direct digital control output channel and the output transducer should be limited to avoid output signal error (see Figure 4-9).

Communication Wiring

Communication wiring systems, also called local area networks (LANs), are signal systems that allow digital information to travel between control devices in the system. The configuration and types of wire used were discussed in Chapter 3. Care must be taken to keep communication wiring away from other wiring systems or devices that generate electric signals that could interfere with the signal transfer of the LAN. An electrical phenomenon known as reactive inductance can cause severe damage to the LAN signal by introducing an electrical opposition to the signal flow. The magnetic field that surrounds an electric conductor, such as the 277-V power wiring serving a lighting ballast, will change the magnetic characteristics of other electric conductors in proximity to it. When a LAN system is near power conductors, this change in magnetism affects the ability of the wiring system to carry the LAN signal and will destroy transmitted data. Therefore, it is imperative that communication wiring systems be installed away from other power and signal systems.

Chapter 5

Interoperable Control Systems

Control System Interoperability

There are three distinct echelons in any digital control system: the control level, the data transfer level, and the user information level. At the *control level,* functional devices containing intelligent microprocessors provide localized control over building systems and equipment. The *data level* is responsible for integrating data from multiple controllers and, using software, adds features to the overall system including trend data collection, alarm management, energy consumption reporting, and the like. The *information level* contains the operator workstations that monitor the activities of the control and data levels and provides a graphical interface between human operators and the network equipment operating at the data level.

In a perfect world, control systems for buildings could be constructed of devices at each of these layers and could be interchanged and added onto to build the optimal system for any given building HVAC application. However, there are constraints that have been perplexing building controls engineers since the advent of microelectronic systems. Although each of the devices at the control and data levels contain embedded microprocessor intelligence, they are isolated from each other by an inability to communicate that is caused by the proprietary

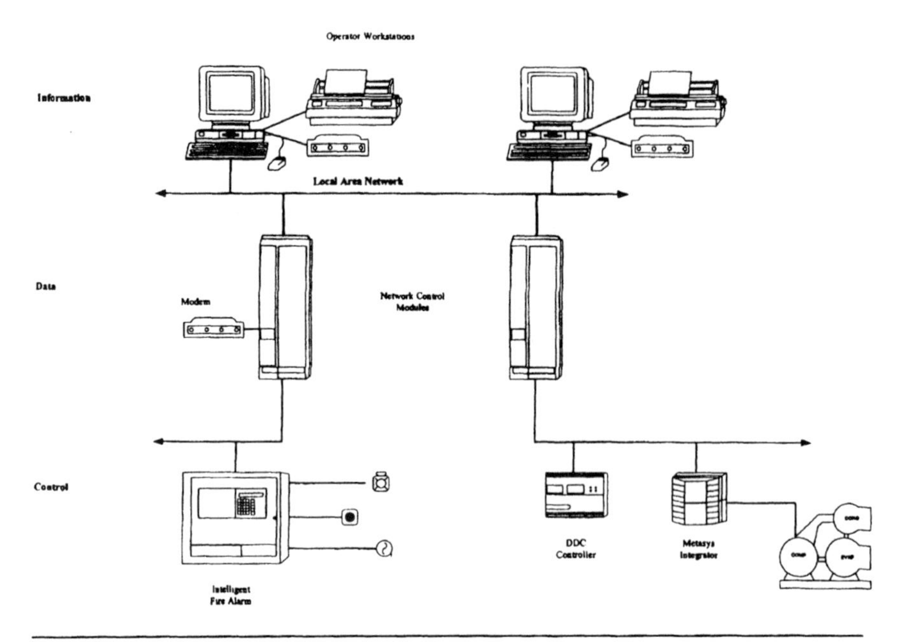

Figure 5-1 Operating Levels of DDC Systems.

Source: Johnson Controls

nature of their design. Direct digital control (DDC) manufacturers design components to work within systems of their own manufacture. This is done to protect their investment in the research and development necessary to develop their systems, and to provide a future market for their products as building owners and operators expand or otherwise change the configuration of their control systems. Since the advent of DDC systems this problem has perplexed building owners, who are often frustrated by the inability of one brand of control device to communicate with another. This prevents the building owner from adding to an existing system devices that serve a future need, and the inability to integrate the operation of these intelligent new devices into the overall configuration of a building control system network only makes the problem worse.

In recent years, demand for open or interoperable systems has grown to such an extent that control manufacturers can no longer ignore it. Although the DDC market today is dominated by suppliers of nonintegrated, proprietary systems, there fortunately exists the underlying technology to support the development of standard configurations and communication protocols for network control systems.

As a result, the natural evolution from closed, customized or otherwise proprietary systems toward the development and proliferation of open, standard, and fully interoperable systems is now underway. The technology required to achieve this interoperability is here today; it is only a matter of giving DDC system manufacturers the economic incentives to embrace this new methodology and to develop systems that will allow more flexibility to end users. Hierarchical proprietary technologies are becoming obsolete as the industry demands much simplified and standardized systems.

There are five distinct levels of interoperability. They are: (1) coexistence (separate but equal); (2) custom built interface (gateways between dissimilar systems); (3) compatible systems (limited communications capability); (4) "plug and play" systems (easily integrated, like stereo components); and (5) fully interchangeable systems.

The difference between these levels of interoperability is defined by *protocols* that are the languages or the message systems that allow these dissimilar devices to communicate.

For many years the means to connect devices from different manufacturers were crude and resided at the control layer of these systems. Interfaced devices such as switching relays and electronic gateways or patches were used to link these systems. Although they met with some success, these methods were slow and problematic, not to mention expensive. Custom-made interfaces between these dissimilar systems were not only difficult to support but expensive as well. Moreover, as building management staff changes took place over time, familiarity with the custom nature of such built-up systems waned and they ultimately fell into disuse and disrepair.

Demand grew for not only interoperable control devices but also for common communication media that would connect information from these devices together into a single system. If this could be achieved, then users could select devices at the control level from among the best available for a given application, as opposed to what was available only within a given product family from a single manufacturer. Parts and subsystems could then be added or subtracted to meet the exact needs of building operators and their facilities. With the downward distribution of microprocessor intelligence being nearly complete, manufacturers are now turning their attention to how these devices can be interconnected. This interconnection will come from the development of a common protocol or communication language that will make it possible for all of these devices to communicate with one another.

Today there are more than a dozen protocols competing for domination of the control system market. Although it is beyond the scope of this chapter to address them all, we will focus on two emerging technologies that show the most promise of being adopted as enduring standards. They are the building automation control network (BACnet) and the LonWorks® technology that has been developed by Echelon

Corporation. These are two distinct methodologies for facilitating digital communication between systems of dissimilar manufacture which differ significantly. The BACnet protocol is primarily a communication standard used to integrate systems at the data level. The LonWorks technology involves embedding neuron chips in control devices at the control layer, which will give them the power to intercommunicate as well as to operate at the local level.

BACnet

In response to demand from building users for a nonproprietary communication protocol, the American Society of Heating, Refrigeration and Air Conditioning Engineers (ASHRAE) formed Standards Committee SPC-135P, whose objective was to define the perimeters for a standardized communication protocol, also known as a communication backbone, that would support multiple vendor building automation system products on a single network. The committee's purpose was to identify the needs of the HVAC industry relative to the integration of multiple DDC systems and define how the needs of the HVAC industry will be addressed through the standard. The committee was then charged with evaluating and employing industry standard technologies wherever possible to allow all manufacturers to implement this standard competitively. The ultimate result of this effort was the delivery of a standard that would allow DDC manufacturers to deliver communication compliant systems without hindering creativity or market differentiable features that were so important to private industry.

The committee derived the term *BACnet* from the phrase *B*uilding *A*utomation and *C*ontrol *net*work. This standard establishes a set of hardware and software rules that include the methods for connectivity, communication protocols, services, and data objects within the standard. It created a guideline that manufacturers could follow, broken down by degrees of connectivity and user interoperability, that would allow independent DDC systems to operate together on a common network. At its essence, this protocol preserves the uniqueness of proprietary systems, while providing a means for them to communicate above the device level using a common language.

The BACnet protocol is based on the OSI seven-layer model described on page 75. However, it does not employ all layers, only those that promote interoperability. To become compliant with this standard every manufacturer will have to meet four basic implementation steps: (1) hardware interface (physical layer), (2) communication services interface, (3) BACnet objects level, and (4) BACnet object services level. By complying with the four implementation levels, a DDC product can become BACnet compliant. However, there is a final implementation level that all manufacturers must complete which is the delivery of a BACnet gateway that will allow the current proprietary network and protocol to coexist with other BACnet-compliant products. The interface devices and application specific controllers at the control layer will all be affected in one form or another by BACnet. The degree of this compliance is not proscribed or governed, and it is up to individual manufacturers to determine the extent to which the standard is applied to their particular products.

There are six levels of conformance with the BACnet standard. Each conformance level corresponds to an increase in the quantity and complexity of the objects and services supported by a BACnet compliant controller or control device. The idea is that more complex conformance classes will be applied to higher level controllers where memory and microprocessor speeds are not an issue. In essence, these conformance classes allow manufacturers to scale the degree of BACnet protocol implementation. Within this scaling paradigm are 12 functional groups that help manufacturers define such user-sensitive functions as how the hand-held interface is to be designed and operated, how the user will interface through a PC using software that communicates with the overall system, how events are initiated from within a system across products of different manufacturers, and how data is collected and stored. When specifying systems applying BACnet standards it is absolutely critical that these levels of BACnet connectivity be carefully specified; otherwise manufacturers will not know to what degree they are expected to comply with the standard. More information on how to specify a BACnet-compliant system is provided later.

Once a standard such as BACnet is created, the issue of how to certify manufacturers' compliance and verify their adherence to the standard comes into question. The ASHRAE committee decided that is not within their scope to address such enforcement issues and turned elsewhere for assistance. The National Institute for Standards Technology, also known by the acronym NIST, sponsored an inter-operability consortium comprised of most of the major DDC manufacturers. The objectives of the consortium were to evaluate and prove the viability of the BACnet standard and to provide a methodology for compliance testing before certifying products as "BACnet compliant."

Echelon

Echelon Corporation is a privately held business in Palo Alto, California that was founded in 1988 for the purpose of developing and marketing a form of microprocessor known as a *neuron chip.* A neuron chip is an integrated circuit that incorporates communications, control, scheduling, and input/output functions in its firmware. Echelon arranged to have Motorola and Toshiba manufacture these chips to their specifications to ensure adequate supply and world class quality. Echelon then configured these chips to work within a local operating network or LON which is further comprised of nodes, or intelligent control devices, that communicate using a common protocol. By linking intelligent control devices that have embedded communication technology, Echelon created a multilayered protocol that would support an open architecture. In other words, any controls manufacturer that uses Echelon neuron chips in its products makes it possible for its product to communicate with any other products using the same chip. Whereas BACnet addresses a communication standard at a network or horizontal level, Echelon's approach has been to introduce interoperability into systems from a device-oriented or vertical level.

The Echelon protocol is a complete seven-layer communications protocol that also allows nodes or devices to interoperate using an efficient and reliable communication language that is a simple exchange of control information. The interface between devices via the network is provided by shared data items called *network variables*. These network variables are arranged into standard types also known as standard network variable types (SNVTs or "snivvets") that allow devices to exchange data in an efficient manner. This is ideal for devices such as sensors and actuators in that it allows "plug and play" compatibility on a common system.

Most manufacturers today that are utilizing the Echelon technology are incorporating SNVTs into their product design to allow their devices to communicate on this open protocol. Although many manufacturers have embraced the Echelon technology, there are differences among these manufacturers regarding the level of Echelon features they have implemented into their designs. Therefore, full interoperability and a totally open system has not yet been achieved.

Interoperability offers extreme advantages to an end user. It allows competitive bidding by commoditizing control systems and it avoids the increased cost and dependency involved in utilizing proprietary systems. Moreover, future expansion without replacement of the existing system is possible and there is more flexibility in the configuration of the overall control system. Obviously, interoperability is a threat to manufacturers of older proprietary systems. However, today most DDC manufacturers are viewing interoperability as a new market opportunity and are rushing to embrace these open systems.

Chapter 6

Direct Digital Control Application Strategies

This chapter addresses the application of direct digital control (DDC) systems to common HVAC control situations and focuses on how DDC can be used to enhance HVAC system performance. The intent is to apply the power of digital control systems to create "intelligent" HVAC systems that can automatically adapt to environmental and system changes. In addition to system operating improvements, a strong emphasis toward energy conservation is incorporated into the control designs.

Local Control Strategies

The control strategies described in the following sections are proven for reducing energy consumption while maintaining occupant comfort. Although these strategies traditionally have been accomplished with conventional controls, the unique powers of DDC enhance the effectiveness of these control methods by improving the speed of the control system response and by providing accurate data on HVAC system performance. To help the reader think in terms of digital control, a point table is included for every control strategy presented.

Minimum-Outside-Air (Ventilation) Control

The purpose of outside-air control is to maintain the minimum quantity of outside air required to satisfy the ventilation requirements of a building. In addition, outside-air control is used to take advantage of the free cooling available in outside air to satisfy demand before initiating mechanical cooling processes. With the growing problem of indoor-air contaminants and the related impact on health and worker productivity, engineers are being called upon to address this problem in their designs. At the industry level, the American Society of Heating, Refrigeration and Air Conditioning Engineers (ASHRAE) has recently increased the minimum requirement for ventilation air from 10% to 15% of the total system volume.

There are two methods for ensuring that minimum ventilation air control is accomplished: pressure-dependent minimum outside air and pressure-independent minimum outside air. These strategies are similar in design to the controls used on VAV terminals.

Pressure-dependent minimum-outside-air control is the most common method of providing ventilation air to a building. This method uses an outside-air control damper with a mechanical linkage set to prevent the damper from completely closing. The minimum open area of the damper is mechanically set when the air system is balanced, based on the minimum ventilation requirements of the building. However, there is a fallacy in this method of control when it is applied to variable-air-volume (VAV) systems. In VAV systems, the speed of the supply fan is modulated based on the demand for cool air at the occupied zones. When the fan speed is reduced, a severe drop in the volume of ventilation air at the outside-air damper will occur because of the related drop in the velocity pressure of the system. From the fan speed law we know that velocity pressure varies at a squared rate to the fan speed. Therefore, when the supply fan is operating at 50% of maximum, the velocity pressure of the system is at 25% of maximum. The quantity of outside air falls far below the desired minimum under such conditions. This condition can create significant problems in buildings where smoking is allowed or where hazardous materials are being handled. An improvement on the pressure-dependent minimum-outside-air strategy is to arrange a control scheme that guarantees a minimum volume of ventilation regardless of system fan speed. Control sequences that compensate for changes in system pressure are known as pressure-independent.

The *pressure-independent minimum-outside-air control* loop modulates the position of the outside-air damper to ensure that a minimum volume of outside air enters the HVAC system regardless of the speed at which the fan is operating. Figures 6-1a and 6-1b illustrate the

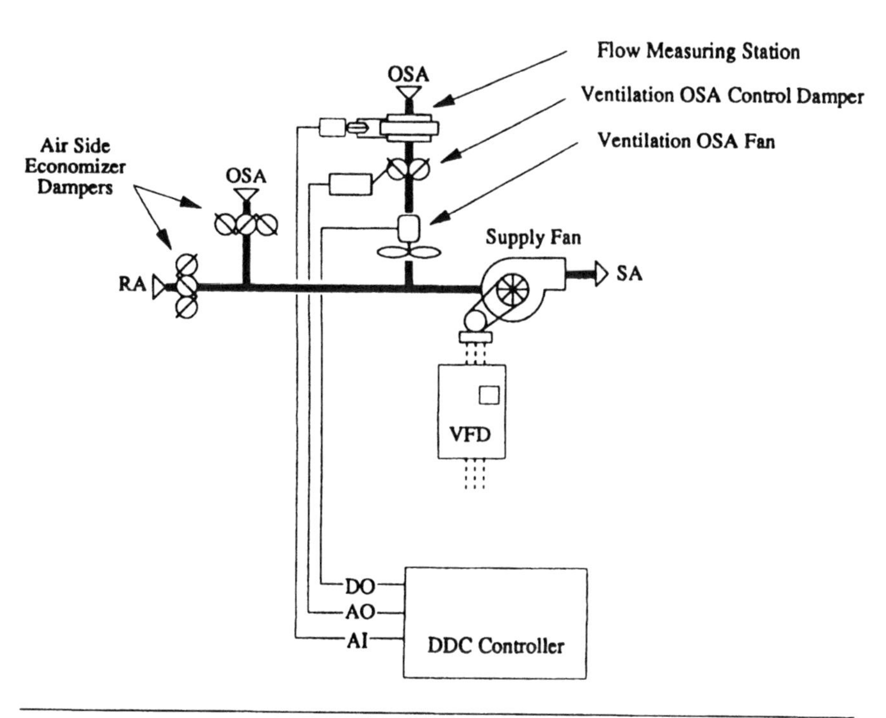

Figure 6-1a Minimum outside air for ventilation.

hardware required for this system. An airflow measuring station in a parallel outside-air duct measures the quantity of air passing through its outside-air damper at all times. With the actual quantity of the outside air known, the direct digital controller modulates the outside-air damper to ensure that the desired minimum quantity of outside air is maintained regardless of the static pressure in the mixing plenum.

When fan speed is reduced, the reduction of system pressure in the mixing plenum creates another control problem. An airflow measuring station, like any other flow-sensing device, requires a minimum pressure drop across its apparatus to measure volume accurately. Reduced fan speeds result in air velocities that fall below the minimum control range of the airflow device. Some airflow measuring (AFMS) devices have been designed to measure at very low velocities; however, at such low velocities, a slight change in pressure can have a large impact on the control signal. Therefore, it is preferable to provide sufficient velocity so that a stable control signal is possible.

To solve this problem, a small in-line fan is used, as shown in Figure 6-1a. This fan provides a constant velocity to allow the AFMS to measure accurately, and it prevents mixing plenum air from exhausting

Point Description	(DO)					(AO)						(DI)				(AI)						
	Start/Stop	Open/Close	Other	Alarm		Position Control	Speed Control	Control Point Adjust	Other	Alarm		Status	Other	Alarm		Temperature	Pressure	Flow	Speed	Volume	Other	Alarm
Outside Air Volume (CFM)																				X		
Start/Stop Command, OSA Fan	X																					
Position Control, OSA Damper						X																

Figure 6-1b Input/output checksheet: minimum outside air control.

through the outside-air ducts when the economizer system is in the minimum-outside-air mode. Minimum outside air is provided under all operating sequences by simply modulating the position of the control damper. This method is highly accurate, but it does introduce an increase in the first cost of the system because additional duct work, an airflow station, and an extra fan are required. For environments where minimum ventilation is crucial, this additional cost is easily offset by the benefits of controlled ventilation.

Mixed-Air Control

Mixed-air control, often referred to as economizer control, is frequently used in dry climates with low dew-point temperatures. Economizer control takes advantage of the cooling or heating energy available in outdoor air. Outdoor-air and building return-air temperatures are mixed in proportions that will satisfy the mixed-air temperature set point while the least amount of mechanical heating or cooling is used. Sensors must be located in the outside-air, the mixed-air, and the return-air ducts so that all three temperatures can be measured simultaneously.

Mixed-air control is closely related to minimum-outside-air control in that these two strategies must operate in concert to provide a guaranteed minimum volume of outside air to the building while capitalizing on the most attractive source of available heating or cooling energy in the incoming airstreams.

Figures 6-2a and 6-2b illustrate an air-handling system utilizing both economizer outside-air and ventilation outside-air ducts. The point of control for mixed-air temperature is downstream of the filter bank and upstream of the heating and cooling coils. This location provides a well mixed air temperature from the mixing plenum, and an averaging type sensor is used to determine an accurate representative temperature. The DDC controller modulates the outside-air and return-air dampers as needed to maintain the mixed-air temperature set point. When the outside-air temperature exceeds a preset maximum, the economizer outside-air damper closes to use as much return air as possible to reduce the mechanical cooling demand on the system. The minimum-outside-air damper continues to modulate as needed to maintain the desired minimum-ventilation-air volume. When the system fans are off, the outside and exhaust dampers are closed, the return-air damper fully opens, and the minimum-outside-air damper continues to modulate.

By adding humidity sensors in the economizer outside-air and return-air ducts, one can measure the total heat content of the airstreams, thus allowing enthalpy control to be achieved. Enthalpy con-

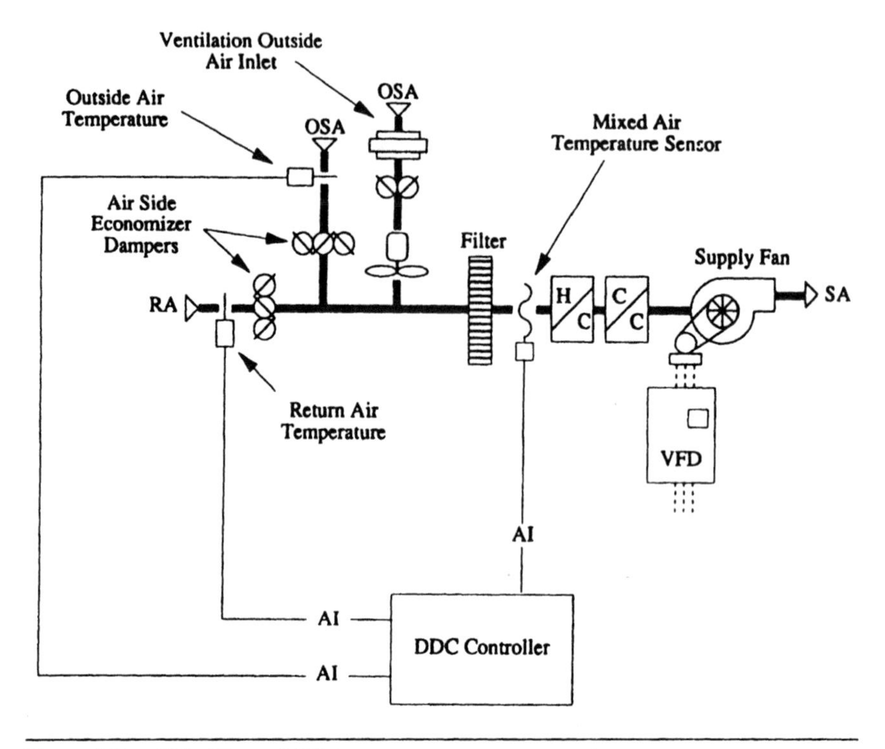

Figure 6-2a OSA, RA, and MA temperature measurement.

trol is used in humid climates to select airstreams based on total heat content of incoming air. Enthalpy control techniques are discussed later in this chapter.

Static-Pressure Control

The purpose of *static-pressure control* in VAV systems is to maintain a minimum amount of static pressure in the duct system to support the proper operation of air distribution devices. The two primary modes of static-pressure control are *duct static-pressure control* and *building static-pressure control.* Duct static-pressure control is accomplished by modulating the volume of air coming from the supply fan as measured by a static-pressure-sensing tip downstream of the fan in the duct system. Building static-pressure control is accomplished by means of a static-pressure-sensing tip located in the occupied space. This signal is transmitted back to the DDC panel, and the volume of the return fan is modulated to maintain the desired building static pressure.

There is no direct relationship between the duct and building static-pressure control systems; each acts independently to satisfy its respec-

	(DO)					(AO)						(DI)				(AI)						
Point Description	Start/Stop	Open/Close	Other	Alarm		Position Control	Speed Control	Control Point Adjust	Other	Alarm		Status	Other	Alarm		Temperature	Pressure	Flow	Speed	Volume	Other	Alarm
Outside Air Temperature (°F)																X						
Return Air Temperature (°F)																X						
Mixed Air Temperature (°F)																X						

Figure 6-2b Input/output checksheet: outside-, mixed-, and return-air temperature.

tive set points. However, when these two independent systems operate concurrently, they indirectly affect each other. As changes in building loads cause changes in the delivered air volume to occupied spaces, the supply fan slows down, causing a change in the space pressure, which in turn causes the return fan to slow down. When building loads are high, the opposite sequence occurs.

There are many methods for controlling air volume in HVAC fan systems. The most common are discharge dampers, inlet vanes, and variable-frequency inverters.

Discharge dampers are simply control dampers installed at the outlet of a fan. The discharge damper modulates open or closed in response to changes in system pressure. Discharge dampers provide excellent control performance by modulating to allow only enough air to maintain system pressurization; however, their operating efficiency is very poor because of the large amount of energy consumed by the fan when it is operating against a flow restriction.

An *inlet vane damper* modulates the open area of the fan inlet on centrifugal fans by means of a scroll damper linked to an actuator. By proportioning the amount of inlet air, this control method modifies the fan operating curve to change the output volume of the fan. This method requires less horsepower than the discharge damper method by efficiently approaching the fan curve. Inlet vanes, unfortunately, require a high degree of maintenance and can generate a lot of acoustic noise.

Variable-frequency inverters, or variable-speed drives, are an efficient method for controlling electric motor speed by modulating the voltage and frequency of the motor power supply. Unlike discharge dampers and inlet vane dampers, which impose a mechanical resistance on the flow of the fan, a variable-frequency drive modulates the output of a fan without affecting the operating efficiency of the fan motor. Although independent control of duct static pressure and building static pressure is acceptable in most commercial office environments, these methods do not provide an adequate level of pressure control in critical environments, such as hospitals, laboratories, and research centers. When critical control of space pressures is required, volumetric control strategies should be considered. Volumetric control is achieved by measuring the quantity of air moving through the supply, return, and exhaust ducts of a system. If air quantity is expressed in terms of cubic feet per minute (CFM), a positive, neutral, or negative pressure can be accurately maintained in virtually any space.

An airflow measuring station is used to calculate actual CFM from the velocity pressure and cross-sectional area of the duct. The AFMS provides an analog input to the DDC panel, which will modulate the speed of the supply fan and return fan and adjust the exhaust-air damper to maintain the desired volume relationship of the system.

An infrequently used method of HVAC system pressure control is known as fan tracking, in which the supply and return fans are operated in parallel based on an arbitrary relationship between duct static and building pressures. Accurate fan tracking is difficult to achieve because it relies on the similarities in the fan curve between two individual fans that are of different size and operating characteristics. In practice, these curves are usually difficult to parallel, resulting in inaccurate control.

A recommended strategy for achieving accurate pressure control using DDC is illustrated in Figures 6-3a and 6-3b. In this arrangement, an airflow measuring device is installed in the supply- and return-air ducts, and a static-pressure sensor is located in the supply duct to the VAV terminals. Note that this arrangement includes a pressure-independent minimum-outside-air control using a control damper and an in-line fan. Variable-frequency drives are used to modulate the speed of the fans. The supply-fan speed is modulated as required to maintain the duct static-pressure set point while the return-fan speed

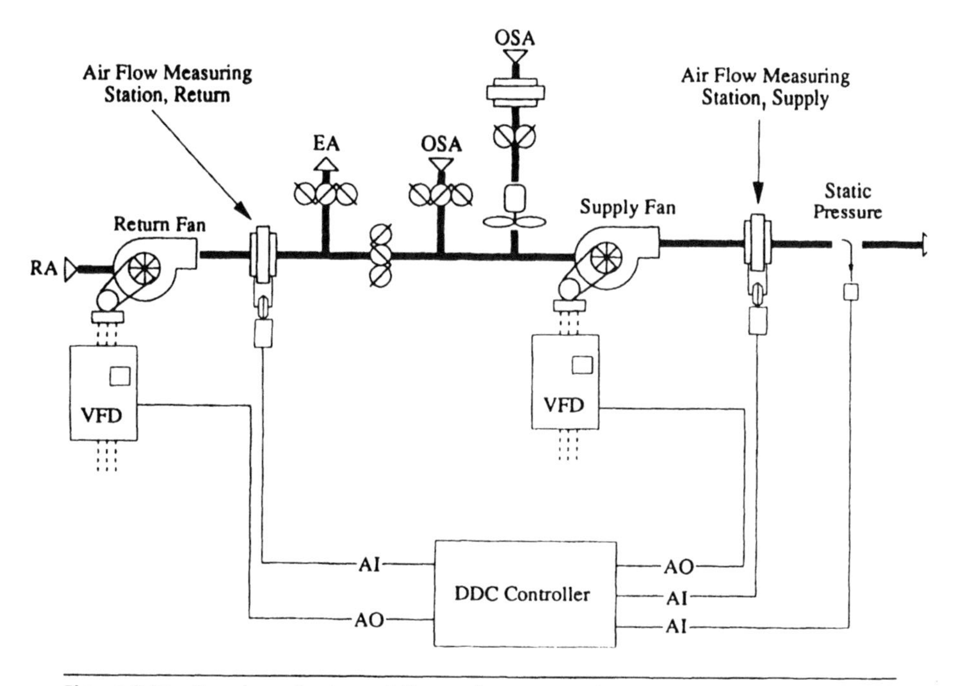

Figure 6-3a Supply and return airflow volume control.

	(DO)					(AO)						(DI)				(AI)						
Point Description	Start/Stop	Open/Close	Other	Alarm		Position Control	Speed Control	Control Point Adjust	Other	Alarm		Status	Other	Alarm		Temperature	Pressure	Flow	Speed	Volume	Other	Alarm
Supply Air Volume (CFM)																				X		
Return Air Volume (CFM)																				X		
Supply Fan Speed Control							X															
Return Fan Speed Control							X															

Figure 6-3b Input/output checklist: supply and return airflow volume control.

is modulated to maintain the desired volumetric differential between the supply and return fans.

The following is an innovative (albeit aggressive) variation on this strategy that capitalizes on the power of DDC systems: Reset the duct static-pressure set point based on VAV terminal damper position on the theory that the terminals do not need as much static pressure to operate when HVAC loads are low. The positions of the VAV terminal dampers indicate the load at the zone level, which resets the duct static-pressure set point. This in turn controls the speed of the supply fan. The zone damper that is open the farthest determines the rate of static-pressure reset because it is assumed that it represents the zone with the greatest cooling demand. As the duct static-pressure set point is lowered, all of the terminals will gradually open more to maintain their zone set points. In theory, this should improve the accuracy of the zone control because the terminals will be operating within the most accurate range of their stroke. To accomplish this strategy, one needs intelligent VAV terminal controllers that can provide damper position values to the control program, and one should carefully evaluate the duct system and terminal arrangements to make sure that this strategy is applicable. When arranged properly, this strategy can substantially reduce fan speed during the mild weather conditions encountered during spring and early fall.

Variable-Air-Volume Terminal Control

The purpose of VAV terminal control is to modulate the cooling supply-air temperature to meet the minimum cooling requirement of the space. Space temperature sensing and terminal control traditionally have been accomplished with pneumatic, electric, and electronic devices.

The most common form of VAV terminal control is pneumatic. In this arrangement, a combination sensor/controller thermostat controls a pneumatic actuator on the damper shaft of the VAV terminal. As space temperature changes in response to changing load conditions, the space thermostat modulates the terminal damper to allow more or less cooling air to enter the space as required to maintain the zone set point temperature.

The volume and velocity of air entering the space through the VAV terminal are determined by the pressure characteristics of the duct distribution system. In environments where fluctuations in air volume are not critical to occupant comfort or manufacturing processes, a pressure-dependent terminal can be used. Pressure-dependent terminals will modulate air volume in response to changes in space temperature irrespective of upstream static pressure. In systems where a minimum quantity of supply air must be provided under all conditions, a pressure-independent terminal is required. In this arrangement, an averaging type pressure sensor is introduced just upstream of the VAV terminal to provide an input signal to the controller of the income static pressure. The controller can then convert this information into a measurement of volume and can override the temperature control of the terminal damper to maintain a minimum CFM. Pressure-independent VAV terminals are used frequently in variable-volume fan systems. As individual pressure-independent terminals vary the quantity of air to their zones, the pressure conditions in the supply duct are in a state of constant change. Through the duct static control strategies described earlier, the speed of the supply fan can be modulated to maintain only enough static pressure to allow the VAV terminals to control properly, thus saving considerable energy in reduced fan speeds.

The advantages of pneumatic VAV terminal control are that it is a proven method, it is fairly inexpensive, and it is easy to understand. Its disadvantages are that individual zone information is not available at a central point, the minimum and maximum CFM settings of each terminal cannot be reset from a remote location, the temperature set point of each VAV terminal must be set individually at the zone thermostat, and set points have a tendency to drift out of calibration over time.

Control products manufacturers are now marketing small local-loop controllers for VAV terminals as a direct replacement for traditional pneumatic control components. The microprocessor-based VAV controller can monitor and control the temperature of supply air, zone, and air velocity, and can measure and monitor the VAV damper position as an indication of cooling demand. Furthermore, the minimum and maximum settings of each terminal can be remotely reset and monitored. This provides a significant advantage for owners of multitenant office space where changes in occupancy may require a change in the capacities and set points of terminal units installed in concealed or otherwise inaccessible areas.

For VAV systems using terminal reheat, a proportional electronic reheat valve is added to the system. Figure 6-4 illustrates a pressure-independent VAV terminal with a reheat coil and control valve. The control sequence for this arrangement starts with the terminal in the full cooling mode at its maximum position during the periods of high cooling load. As the space temperature decreases, the quantity of supply air decreases until the VAV terminal is operating at its controlled CFM. If the space temperature continues to drop, the terminal controller will slowly modulate the reheat valve to increase the flow of hot

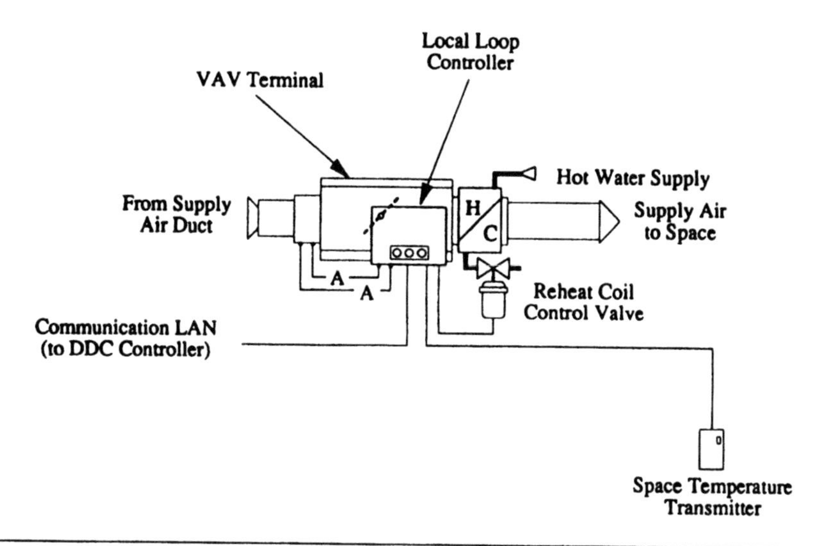

Figure 6-4 DDC air terminal control.

water through the valve as needed to maintain the space temperature after the terminal damper is controlling at its minimum position.

The most significant benefit of microprocessor-based terminal control is information sharing over the entire network system. The collection and evaluation of information on space temperatures and the position of controllers are critical to the accurate performance of control strategies such as supply-air reset. In the case of supply-air reset, knowledge of each zone load in the building allows the local-loop controller to reset the supply-air temperature to satisfy the greatest cooling or heating demand in a building. A concern stated earlier regarding supply-air reset control dealt with the selection and application of representative zone temperatures. In a networked DDC terminal system, all temperatures are available for polling by the system.

Supply-Air Temperature Reset Control

The purpose of supply-air reset control is to reset the temperature of the supply air when there is a reduced demand for cooling. Temperature reset control techniques are almost always used in VAV systems, especially in states that require adherence to mandatory energy saving programs. In California, for instance, Title 24 of the California Energy Commission requires the reset of supply-air temperature in cooling applications as a means of reducing the energy used to produce mechanical cooling and to reduce the energy consumed by a fan.

Unlike heating demand, cooling demand cannot be accurately anticipated by observing outdoor-air temperatures. Closed-loop feedback is required to tell the control system what the true cooling load is in the occupied space. Because building loads are distributed and diverse, the temperature information the DDC system uses to modulate supply-air temperature must be gathered from selected zones as inputs to the DDC panel. The selection of these zones, known as reset zones or sampled zones, presents two problems for the design engineer:

1. How many zones should be sampled?
2. Which zones should be sampled?

In systems utilizing pneumatic zone terminals, signals from selected VAV terminals, expressed as a percentage of load, must be transmitted through a network of high-signal selectors into the DDC panel to indicate the load from the zone calling for the most cooling. Once this load is known, the supply-air temperature may be reset to satisfy that zone. Usually, supply-air reset is modulated over a 10° range, from 55 to 65°F.

Pneumatic high-signal selection requires that branch control signals be collected in a high-signal selector device whose output signal is then returned to the DDC panel through a transducer. This method is expensive in terms of first-cost and presents a substantial maintenance problem. The most accurate means of supply-air reset control is achieved by monitoring intelligent zone terminals that transmit electronic information to the DDC panel over a local area network (LAN). Because the information in a digital zone terminal controller can be shared with other digital controllers, every zone in the HVAC system can be polled for information concerning the temperature in any space, the position of any terminal damper, and the overall percentage of system load. The DDC panel can then compare temperature data from zone sensors to determine the actual demand for cooling and reset the supply-air temperature based on this value. Figure 6-5 compares and contrasts conventional and digital methods of accomplishing supply-air reset control.

Enthalpy Control

In climates where a significant degree of moisture is present, measurement of dry-bulb temperature ignores the total heat (enthalpy) of the incoming airstreams. As with economizer control, the use of the airstream with the greatest available energy must be maximized by measuring both the temperature and the humidity of the outside and return airstreams.

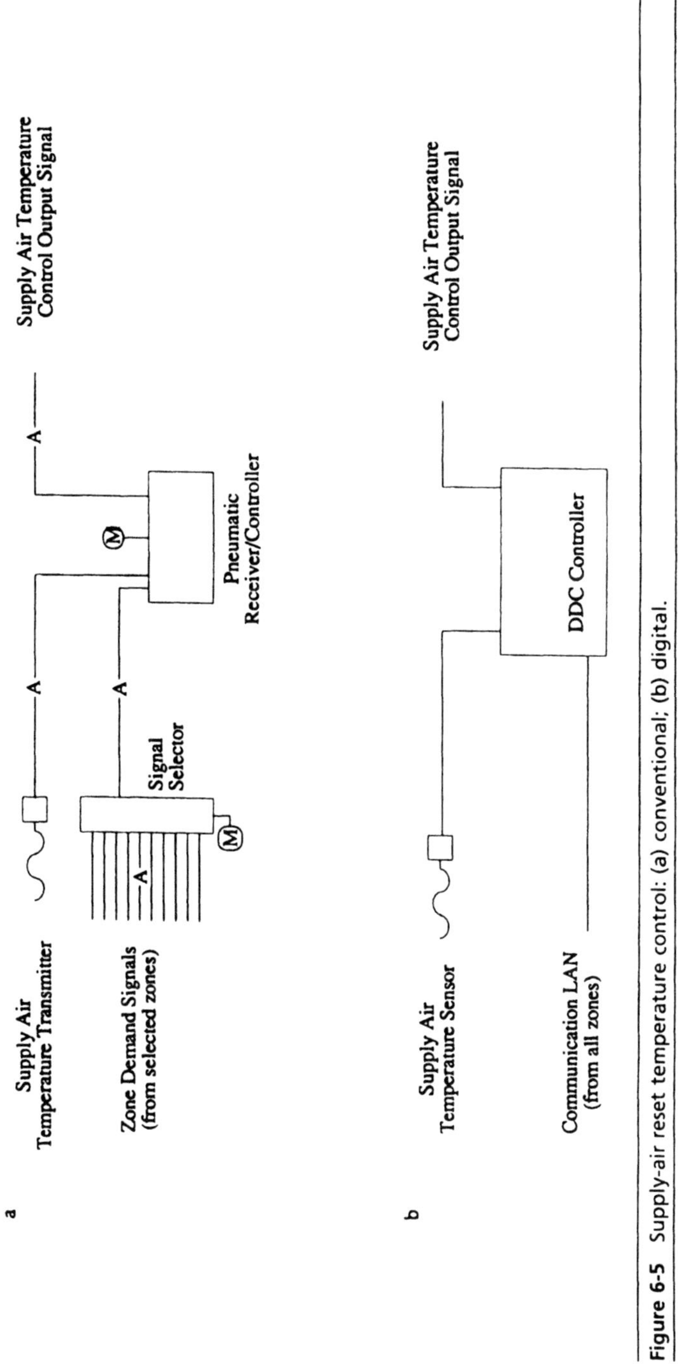

Figure 6-5 Supply-air reset temperature control: (a) conventional; (b) digital.

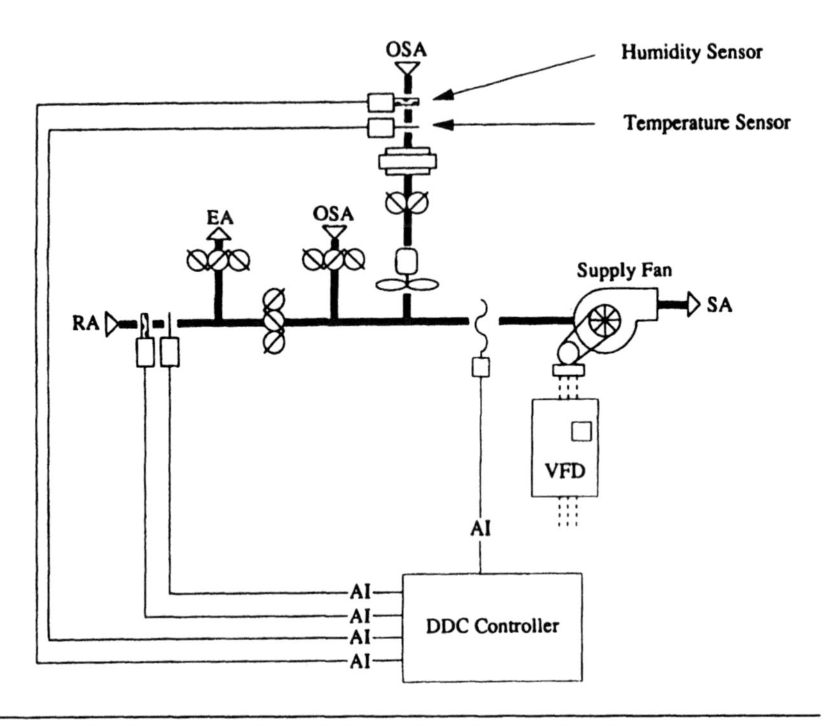

Figure 6-6a Components for enthalpy control of economizer dampers.

In terms of hardware, enthalpy control is identical to traditional economizer control, except that an enthalpy sensor is installed in the outside air and return air in addition to a dry-bulb temperature sensor. Figures 6-6a and 6-6b illustrates the components of an enthalpy control system. Both dry-bulb and wet-bulb temperatures are measured, and the direct digital controller performs calculations to determine which stream is most advantageous in terms of available energy. The controller selects a primary airstream based on enthalpy, then performs economizer control based on the dry-bulb temperature. As with any dry-bulb economizer, an overriding consideration must be given to the outside-air temperature. When it exceeds a predetermined maximum, the outside-air dampers return to their controlled minimum regardless of what the enthalpy of the airstreams may be.

The ability of the direct digital controller to calculate the psychometric properties of air based on wet-bulb and dry-bulb temperature measurements allows sophisticated control to be performed at a relatively low cost. In the past, the cost of control devices and the maintenance of sensitive sensing instruments made the use of enthalpy control

Point Description	(DO) Start/Stop	(DO) Open/Close	(DO) Other	(DO) Alarm		(AO) Position Control	(AO) Speed Control	(AO) Control Point Adjust	(AO) Other	(AO) Alarm		(DI) Status	(DI) Other	(DI) Alarm		(AI) Temperature	(AI) Pressure	(AI) Flow	(AI) Speed	(AI) Volume	(AI) Other	(AI) Alarm
Mixed Air Temperature (°F)																X						
Return Air Temperature (°F)																X						
Outside Air Temperature (°F)																X						
Return Air Humidity (%)																					X	
Outside Air Humidity (%)																					X	

Figure 6-6b Input/output checksheet: enthalpy control of economizer dampers.

marginal at best. With modern high-precision electronic humidity sensors and the cost of DDC system points amortized over the many control strategies being performed in a single system, enthalpy control can now be a cost-effective strategy in humid climates.

Direct Digital Control of Central Plant Systems

Significant opportunities for energy saving exist in most central plant facilities. Several of the most effective applications of DDC to central plant systems are discussed.

Hot-Water Reset Control

A *hot-water reset control* adjusts the temperature of the hot water supplied to building equipment, based on the demand for heat. Demand for heat varies in inverse proportion to outside-air temperature. In conventional control systems, an open-loop strategy is used. Outside air is sensed, hot-water supply temperature is sensed, and the controller modulates a three-way mixing valve to achieve the desired hot-water temperature. Figures 6-7a and 6-7b illustrate this control strategy with a temperature reset schedule based on outside-air temperature. By using a direct digital controller to perform this sequence, a new element can be added to the control scheme: feedback. By placing

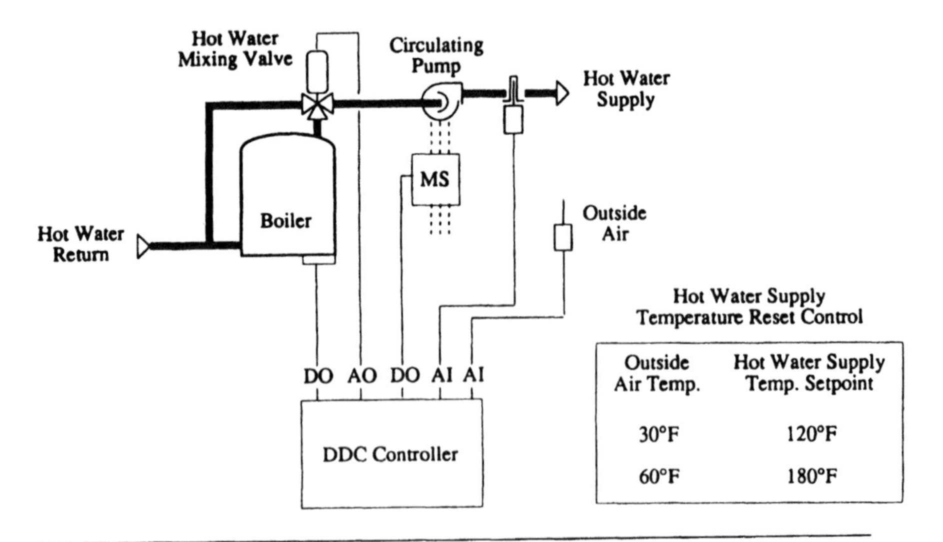

Outside Air Temp.	Hot Water Supply Temp. Setpoint
30°F	120°F
60°F	180°F

Figure 6-7a Heating hot-water reset temperature control.

Point Description	(DO)					(AO)						(DI)				(AI)						
	Start/Stop	Open/Close	Other	Alarm		Position Control	Speed Control	Control Point Adjust	Other	Alarm		Status	Other	Alarm		Temperature	Pressure	Flow	Speed	Volume	Other	Alarm
Hot Water Supply Temperature (°F)																X						
Outside Air Temperature (°F)																X						
Start/Stop Command, Boiler	X																					
Start/Stop Command, Circ. Pump	X																					
HWS Valve Position Control						X																

Figure 6-7b Input/output checksheet: heating hot-water temperature control.

a sensor in the hot-water return line, we can measure an indication of the demand for heating water from the building.

In theory, the temperature of the hot-water supply to the building should be modulated in inverse proportion to the temperature rise across the building load. In practice, however, this does not work, because no consideration is given to individual loads. Figure 6-8 illustrates a simple heating hot-water system supplying hot water to heating coils in VAV reheat zones. Because of the individual loads in the spaces these coils are serving, each coil is at a different position, as determined by monitoring the valve position on the coil. To determine simply the load of the building by measuring the return-water temperature would ignore the fact that some coils are at 100% demand while others are at 20% demand.

A better way to perform hot-water reset control is to look at the position of each hot-water valve in the system. Fortunately, DDC systems make this task easy and affordable. The output control signal driving the hot-water valve can be measured and converted into percentage of valve stroke. Therefore, at any given time, information on what percentage of full flow the valve is operating is available to the DDC panel. By collecting and summarizing these data, the DDC panel can accurately determine the true load in the building and can adjust the hot-water supply temperature upward or downward as needed to satisfy the true demand of the building.

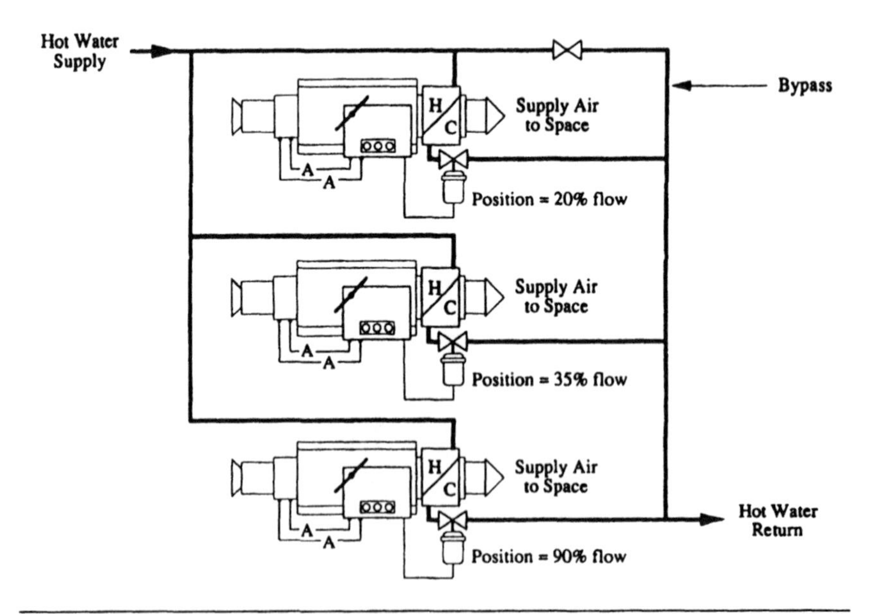

Figure 6-8 Hot-water supply for VAV reheat zones.

Chilled-Water Temperature Reset

Temperature reset of the chilled-water supply (CHWS) temperature can lower the operating cost of chiller plants. An aggressive, yet effective, energy conservation strategy using DDC is to reset the CHWS temperature based on zone valve position. The concept of this strategy is similar to the duct-pressure reset strategy discussed earlier. When the position of each chilled-water coil valve is known, the valve operating under the greatest cooling demand can reset the CHWS temperature upward until this valve reaches 100% of its stroke. As the water temperature gradually rises, the other valves in the system modulate open to satisfy their zone loads, and should operate more efficiently because they are operating near the top end of their stroke. If this strategy is incorporated into the initial design of the chilled-water distribution system, smaller chilled-water pumps and line-sized control valves can be used, yielding a first-cost saving opportunity.

This strategy is, in effect, a means of optimizing chiller performance. However, unlike traditional chiller optimization control, which is based on maintaining a constant condensor-water return temperature, this method monitors the actual building load to control the capacity of the chiller(s), which is more representative of cooling demand than a fixed return temperature set point. Temperature reset strategies can also be applied to the heating hot-water temperature in dual-duct systems using hot-water heating coils. The heating hot-water supply (HHWS) temperature is reset based on building demand, as well as a reference to outside-air temperature as a means of anticipating increased heating loads.

Cooling-Tower Fan Speed Control

One of the most overlooked energy conservation strategies for HVAC systems is the modulation of cooling-tower fan speed in response to changing cooling loads. The principles of cooling-tower fan speed control apply to both direct and indirect cooling-tower designs. Induced-draft or propeller-type cooling towers, as well as forced-draft towers using blower fans, can be controlled by the strategies discussed here. Traditionally, cooling-tower fans run at 100% speed whenever they are operating. Because the motors operating these fans tend to be quite large (from 10 to 50 horsepower), a substantial energy savings opportunity exists. The strategy is quite simple: When demand for condensor-water temperature is reduced as a result of a reduced cooling load, modulate the speed of the tower fan to maintain the condensor-water temperature set point. Figures 6-9a and 6-9b summarize the components of this system. A sensor located in the chilled-

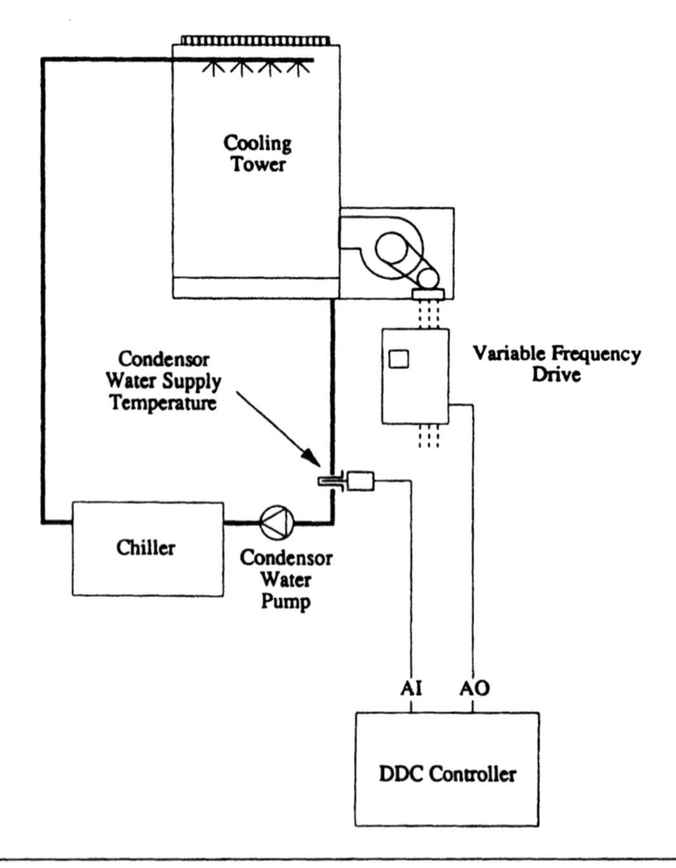

Figure 6-9a Cooling-tower fan speed control.

water supply and a means of modulating the speed of the fan are all that are required.

The methods for modulating the speed of the cooling-tower fan are the same as described earlier for the static-pressure control strategies. Discharge dampers and variable-frequency drives are both used for this purpose.

As with enthalpy control, the effectiveness of this strategy can be enhanced by adding another element to the system: wet-bulb temperature sensing. Adding a wet-bulb temperature sensor to the system allows the chilled-water supply temperature to be tracked along with the wet-bulb temperature so that a relationship can be calculated to determine the ratio between these two values. The theory is simple: the lower the wet-bulb temperature, the easier it is for the cooling tower to maintain its differential temperature and the slower the cooling-tower fan needs to operate to accomplish this differential tem-

Point Description	(DO)					(AO)						(DI)				(AI)						
	Start/Stop	Open/Close	Other	Alarm		Position Control	Speed Control	Control Point Adjust	Other	Alarm		Status	Other	Alarm		Temperature	Pressure	Flow	Speed	Volume	Other	Alarm
Condenser Water Supply Temp. (°F)																X						
Tower Fan Speed Control							X															

Figure 6-9b Input/output checklist: cooling-tower fan speed control.

perature. Modulating the speed of the cooling-tower fan with respect to the change in chiller load and to maintain a constant differential temperature achieves a substantial energy saving. Most simple payback calculations performed for this strategy indicate that the equipment cost will be repaid within 1 year.

The amount of air passing through the cooling tower required to maintain constant condensor water temperature varies with the heat load imposed by the chillers and by the conditions of the entering air at the tower. Therefore, by measuring both the psychrometric properties of the air and the load on the chillers, we can develop a ratio of the actual load to the design load.

The set point of the leaving-water temperature of the cooling tower is set at the condensor water design temperature, which is usually 85°F. It has been shown that more energy will be saved by lowering this set point to 75°F and allowing the chiller to operate more efficiently than can be saved in fan horsepower at the cooling tower. A saving of 1% to 2% of chiller energy per degree of reset can be accomplished and should be used as a magnitude of saving estimate. As the deviation from design set point increases, this percentage of saving decreases radically. Therefore, caution must be used, and it is recommended that this set point not be set below 75°F.

With a DDC system, the building operator can experiment with the condensor-water supply temperature set point to determine the control point at which both the cooling tower and the chiller are operating at an equal rate of efficiency. That is, the operator can make sure that one piece of equipment is not operating to the detriment of others.

Variable Pumping Strategies for Chilled-Water Systems

Variable-air-volume systems were designed to deliver a varying amount of cooling air to a given space, with the speed of the supply fan being modulated as required to supply the minimum amount of cooling air required. Similarly the principles of VAV may be applied to chilled-water pumping systems, thus realizing a potential for energy saving and system simplification. There are two primary configurations for central plant chilled-water systems: single-chiller configurations and multiple-chiller configurations.

Single-chiller configurations utilize one chilled-water machine to supply a constant temperature and quantity of chilled water to the building load. Likewise, *multiple chillers* piped in parallel provide a constant temperature and supply to system loads. Traditionally, both types of systems have been designed using three-way valves to main-

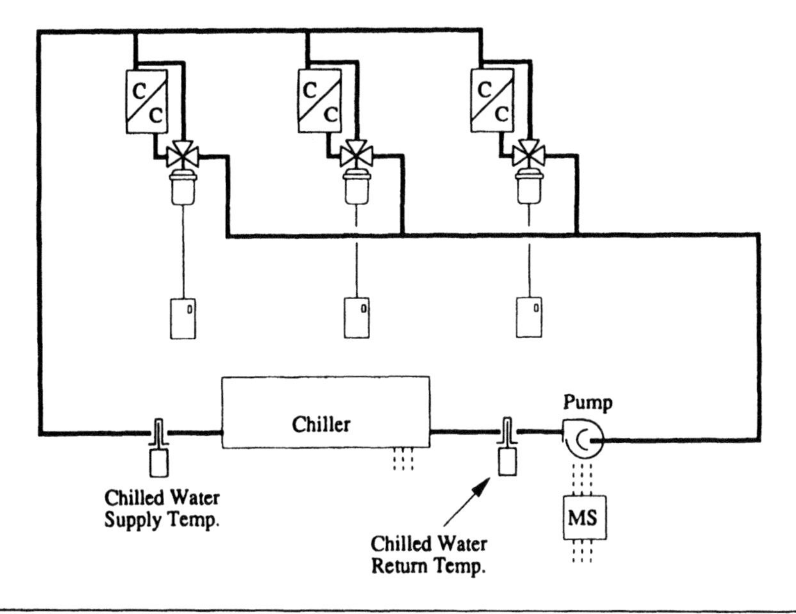

Figure 6-10 Single-chiller configuration, three loads.

tain constant flow through the system regardless of the demands of individual loads. Unfortunately, the recirculation of excess chilled water in these systems consumes a significant amount of energy in system pumping power and chiller operation. This section discusses ways to reduce energy consumption of chillers and pumps with constant-flow loads.

Single-Chiller Configuration

Figure 6-10 illustrates a single-chiller configuration serving several loads. This system uses three-way valves to recirculate excess water to the chiller. The control devices include a chilled-water supply temperature sensor and a return-water temperature sensor. Chilled water is constantly supplied to all loads at a constant temperature. The design engineer has sized each coil at design conditions. However, it is unlikely that all coils will reach peak loads simultaneously. Therefore, this system unavoidably circulates more water than the coils use. This difference between the peak load and the actual load is commonly referred to as *diversity*. In practice, most large chilled-water distribution systems experience very low diversity—commonly 50% to 60%—and this contributes heavily to the energy wasted in recirculating unused chilled water. In fact, in commercial buildings it is very common to find many hours of low system loads while mechanical plants continue to serve the building at peak capacity.

One solution to this problem is to use two-way valves and to vary the flow of chilled water through the system in response to the actual demand for chilled water. Two-way valves eliminate the bypass loop in the system. As individual loads change, the thermostats controlling the valves will modulate the valve to provide only enough chilled water to satisfy the zone demand. Therefore, the flow of the water distribution system will change as load demand changes. In Figures 6-11a and 6-11b a variable-frequency drive for the chilled-water pump motor is added to the arrangement shown in Figure 6-10. Monitoring the position of each control valve in the system allows the speed of the chilled-water pump to be modulated. Although variable control of chilled-water pump speed will yield attractive energy saving, it can cause difficulties with other elements of the chilled-water system. Most chillers do not react well to a change in flow rate, because their coefficient of performance can be drastically altered.

Figures 6-12a and 6-12b offer an arrangement to solve this problem. A bypass loop has been added around the chiller to allow constant flow to the chiller and variable flow to the system. Chiller sequencing

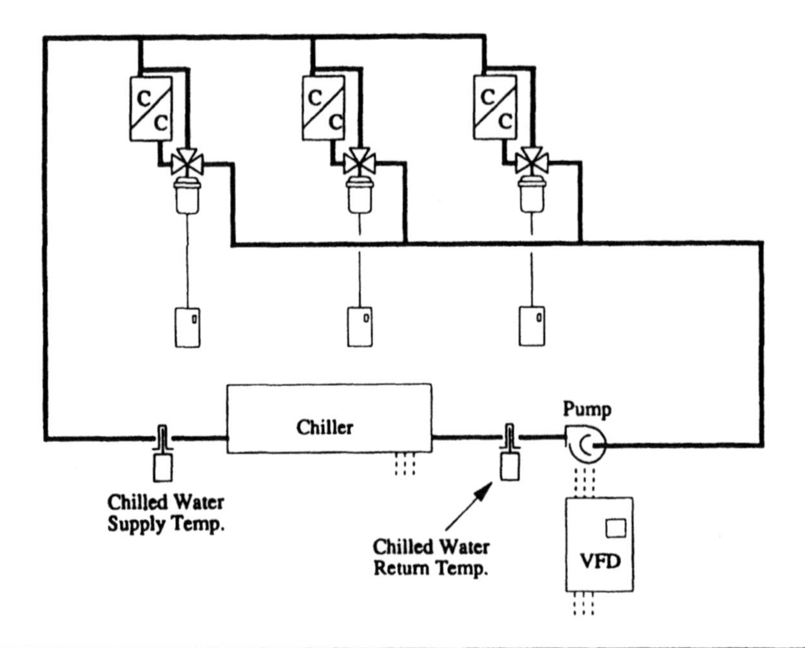

Figure 6-11a Single-chiller configuration, three loads, VFD method.

Point Description	(DO)					(AO)						(DI)				(AI)						
	Start/Stop	Open/Close	Other	Alarm		Position Control	Speed Control	Control Point Adjust	Other	Alarm		Status	Other	Alarm		Temperature	Pressure	Flow	Speed	Volume	Other	Alarm
Chilled Water Supply Temperature (°F)																X						
Chilled Water Return Temperature (°F)																X						
Pump Speed Control							X															

Figure 6-11b Input/output checklist: single-chiller configuration, three loads, VFD method.

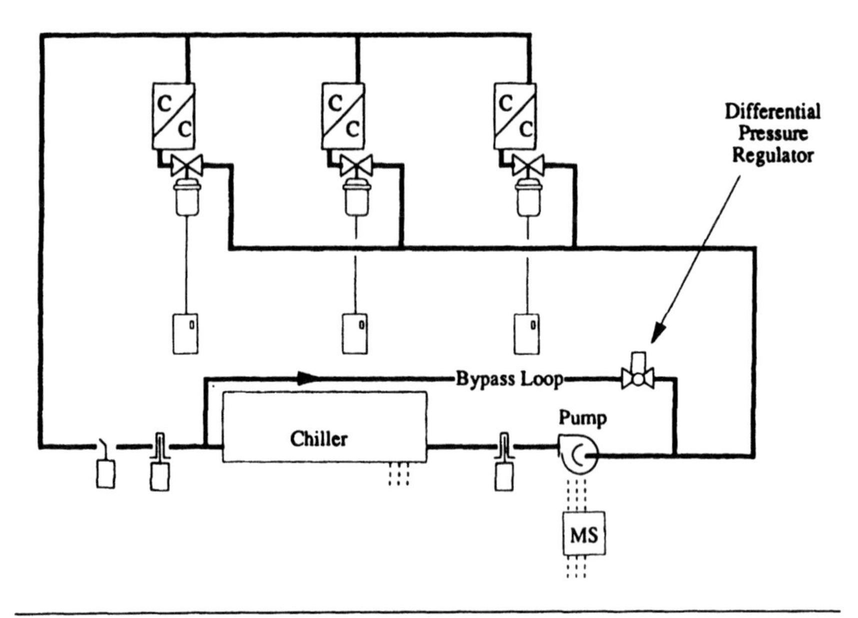

Figure 6-12a Constant chiller flow, variable system flow.

can be accomplished with a return-water temperature sensor and a bypass flow sensor. With this configuration, the staging of the chiller can be initiated by either return-water temperature or supply-water temperature, and because the flow of water through the chiller remains constant the efficiency of the chiller is maintained. This approach, however, does not reduce pumping power because the chilled-water pump must run at 100% of its design speed; the energy conservation opportunity in this arrangement comes from unloading the chillers in response to a continually changing system load. A differential-pressure regulator in the bypass loop maintains a constant differential pressure across the chiller by modulating in response to changes in the position of the two-position control valves. Consideration must be given to the selection of a pump for this type of system to make sure that a constant flow can be maintained under such changing conditions.

Multiple-Chiller Configurations

So far we have evaluated the effect of constant-temperature–variable-flow rates and constant-flow–variable-temperature rates on single-chiller, chilled-water system configurations. In larger central plants, system flow requirements require multiple chillers to satisfy building load. The opportunity for saving energy by simultaneously loading and unloading multiple chillers and for executing variable-flow strategies

Point Description	(DO)					(AO)						(DI)				(AI)						
	Start/Stop	Open/Close	Other	Alarm		Position Control	Speed Control	Control Point Adjust	Other	Alarm		Status	Other	Alarm		Temperature	Pressure	Flow	Speed	Volume	Other	Alarm
Flow Switch												X										
Chilled Water Supply Temperature (°F)																X						
Chilled Water Return Temperature (°F)																X						

Figure 6-12b Input/output checksheet: constant chiller flow, variable system flow.

in such sophisticated systems lends itself well to the use of direct digital controllers.

Figures 6-13a and 6-13b illustrate a multiple-chiller arrangement in a constant-flow–variable-temperature scheme. The only difference between this arrangement and that in Figure 6-12, other than additional chillers, is that an additional temperature sensor is located in the chilled-water return line. The relationship between the temperatures of the return water and the supply water indicates a surplus or deficit of water supply compared with water demand. Measuring the system load as indicated by the position of the control valves enables the relationship between system demand for water and the available supply of chilled water to be staged appropriately to meet this demand. Because this is a constant-flow system, the pump speed remains constant.

The water distribution system arrangements discussed thus far have been based on building loads located close to the central chilled-water plant, so the distribution pumping power is not a matter of great concern. However, in very large or highly distributed facilities, a significant distance between the central chilled-water plant and the building loads is a matter of concern in terms of the pumping power required

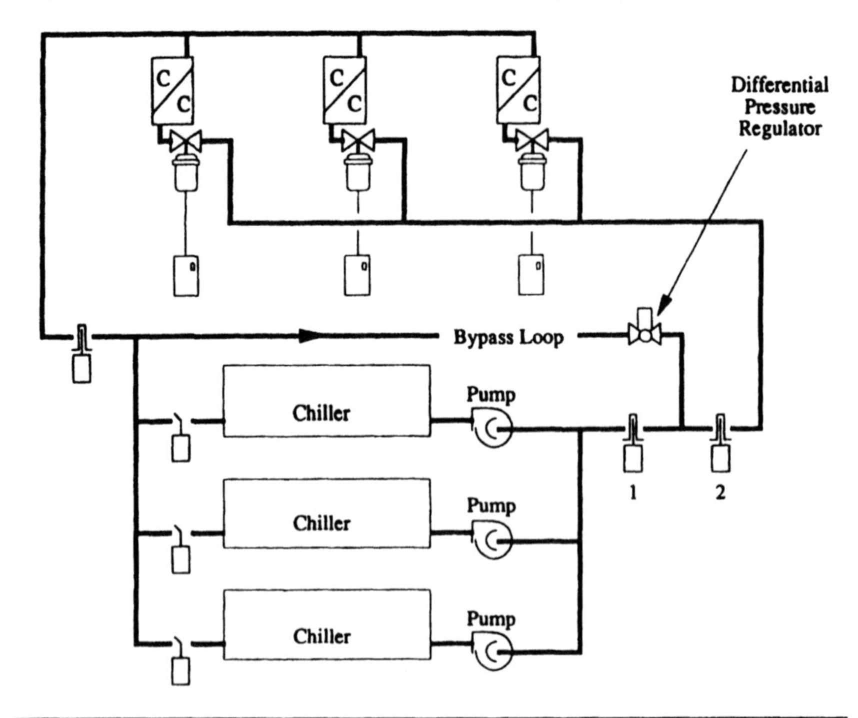

Figure 6-13a Constant chiller flow, variable system flow (multiple chillers).

	(DO)					(AO)						(DI)				(AI)						
Point Description	Start/Stop	Open/Close	Other	Alarm		Position Control	Speed Control	Control Point Adjust	Other	Alarm		Status	Other	Alarm		Temperature	Pressure	Flow	Speed	Volume	Other	Alarm
Flow Switch - Chiller #1												X										
Flow Switch - Chiller #2												X										
Flow Switch - Chiller #3												X										
Chilled Water Supply Temperature (°F)																X						
Chilled Water Return Temp. #1 (°F)																X						
Chilled Water Return Temp. #2 (°F)																X						

Figure 6-13b Input/output checksheet: constant chiller flow, variable system flow (multiple chillers).

to serve the system. By separating or decoupling the chilled-water plant and its distribution network, individual chillers may operate independently under constant-flow conditions while the distribution system requires a variable-flow system that responds to building demand. Figure 6-14 illustrates this concept.

In a decoupler arrangement, a bypass line is added between the chilled-water system and the water distribution system serving the building load. Additional control devices are required to monitor and control this arrangement. A flow switch to indicate the direction of water flow and a flowmeter to indicate the quantity of bypass water are required in addition to the control devices shown. The benefit of

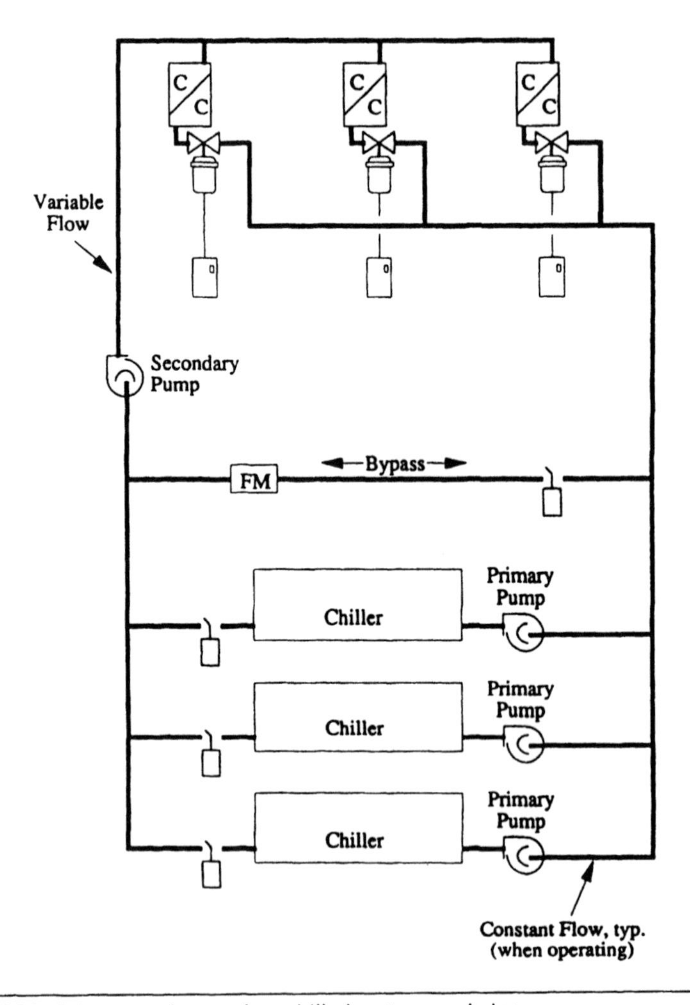

Figure 6-14 Primary and secondary chilled-water supply loops.

this system is that it allows the chilled-water plant to operate under a highly efficient constant-flow–variable-temperature configuration, and it allows the distribution system to the building load to operate in a variable-flow–constant-temperature arrangement, which is most efficient in terms of pumping capacity. Any excess water from the chilled-water plant is automatically returned through the bypass line. This decreases the chilled-water return temperature to the central plant, which then unloads the chillers. When the quantity of bypass water, as measured by the flowmeter, is equal to or greater than the capacity of one chilled-water pump, then that pump and chiller are no longer needed and can be cycled off.

During periods of high cooling demand, the potential for reverse flow through the bypass is possible. The flow switch in the bypass line senses this change of direction and acts through the direct digital controller to initiate additional chillers and chilled-water pumps.

Although multiple-chiller decoupler systems appear complicated, they actually comprise simple control strategies that are combined into one system. DDC systems ease the task of evaluating the performance of such systems by providing graphic displays of the central chilled-water plant and building load distribution piping systems and by providing real-time information on the operating characteristics of the system at various points. Preprogrammed decisions with regard to changes in building load or the direction of flow through the bypass reduce the demand on the building operator's time. Furthermore, aberrations in system operation are quickly detected and an alarm announced by the DDC system. Finally, the ultimate effectiveness of these control strategies can be measured quantitatively by trending sensed data over varying time periods.

Direct Digital Control of Air-Handling Systems

So far, we have discussed the primary strategies to which DDC has been most effectively applied. We now consider these strategies as part of complete HVAC systems, with an emphasis on their interrelationships as subsystems. The systems shown are representative of many systems used in buildings today; however, variations of these arrangements exist, and an understanding of the basic principles of loop control can be applied to virtually any HVAC situation.

Variable-Air-Volume Systems

Variable-air-volume systems were developed during the energy shortage in the early 1970s in response to the needs for a simple air-conditioning system capable of meeting the minimum demands for cooling in a building while reducing energy consumption. Since their introduction, VAV systems have been widely embraced by the building industry because they cost less to install compared with other systems and help reduce overall building operating costs. VAV systems are configured in single-duct and dual-duct arrangements, based on the design of the air distribution systems at the zone level. The problems of control in a VAV system include higher temperature, duct static pressure, and proper operation of the VAV terminals. The components in VAV systems have been described previously and are illustrated in Figure 6-15.

The driving force behind the VAV system is the individual zone temperature transmitter or thermostat. The operation of a VAV system responds directly to changes in zone cooling and heating loads. The sequence of events for a typical VAV system are as follows. When space temperature changes, the VAV terminal controller modulates the terminal damper to maintain the zone set point temperature. When this occurs, a change in the supply duct static pressure causes other VAV terminals in the system to modulate in response to the change in pressure; this change in pressure is sensed by the duct static-pressure sensor, which then modulates the speed of the supply fan to maintain a constant static condition in the duct. The temperature of the supply air to the VAV terminals is modulated based on a determination of the level of demand at the zone requiring the most cooling or heating. The mixed-air temperature control loop, which usually has a fixed set point of 55°F, will modulate the outside-return-and exhaust-air dampers as required to maintain the mixed-air set point. In many cases, however, the mixed-air set point cannot be maintained owing to the effects of minimum outside air or the return temperature from the building. If the mixed-air temperature falls below or exceeds its set point, then mechanical cooling or heating is initiated by modulating water valves on the coils. As evident from this example, the local-loop functions of minimum-outside-air control, mixed-air control, supply reset control, duct static-pressure control, and zone terminal control all act in concert to provide the right quantity and temperature of air to the occupied space. Any change in the load in the occupied space has a ripple effect throughout the system.

Dual-Duct Systems

Dual-duct VAV systems are used when both heating and cooling must be provided to satisfy diverse load conditions in mixed-use buildings. In dual-duct systems, the heating and cooling ducts are arranged in parallel to provide separate streams of air to the zone level where they

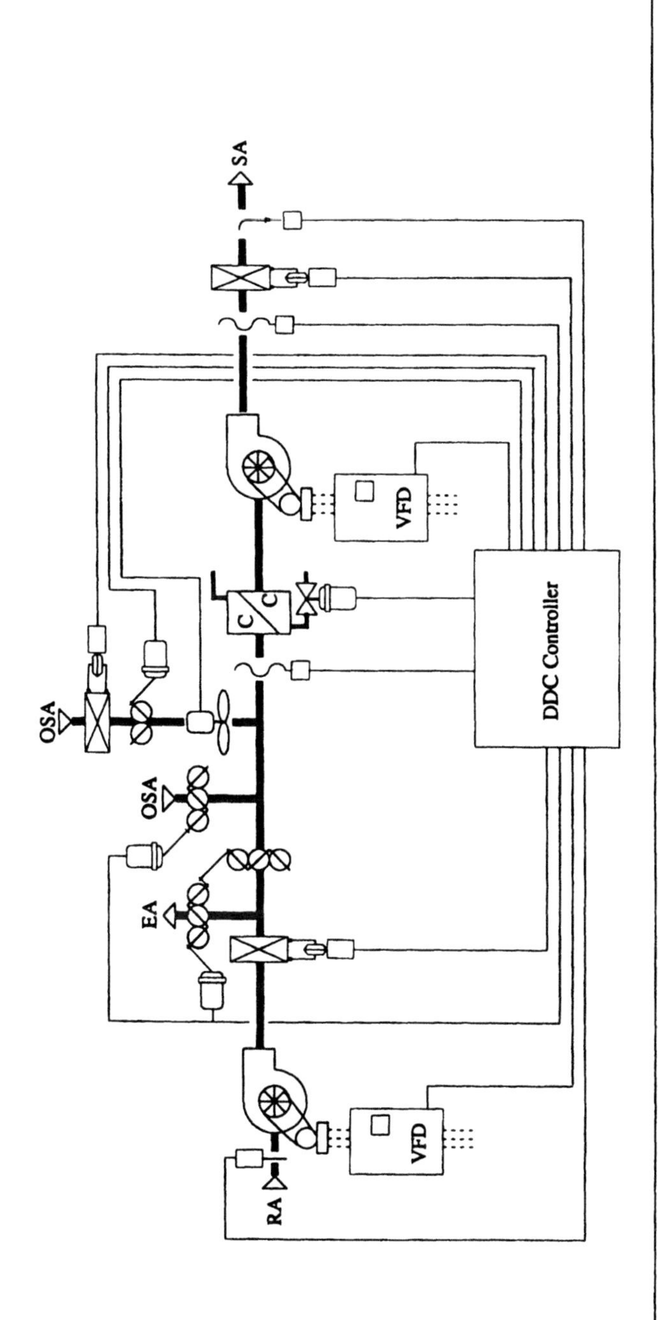

Figure 6-15 HVAC system component schematic diagram (with control devices).

are mixed in a mixing terminal to satisfy the zone temperature set point. Dual-duct systems normally provide a greater degree of comfort to the building occupant, because hot and cold air are mixed to meet the exact needs of the space. Figures 6-16a and 6-16b illustrate a dual-duct variable-volume system utilizing mixing terminals. The application of microprocessor controls to dual-duct systems traditionally is used to convert constant-volume dual-duct systems to variable volume as part as an energy retrofit program. Once converted, these systems operate in a similar fashion to single-duct VAV systems except that the additional component of a control damper for each of the hot and cold supply ducts must be added to perform proper static-pressure control of each air distribution duct. Each control damper is modulated in response to changes in the damper positions at the VAV terminals; the supply fan volume is modulated in response to changes in the hot- and cold-deck control damper positions. Dual-duct systems can be arranged with a single supply fan or with a supply fan and return fan. The distribution of supply air to the zone terminals remains the same. Figure 6-17 illustrates a dual-fan dual-duct system showing the location of control components.

Multizone Systems

Multizone systems are similar to single-zone VAV systems except that heating and cooling coils are arranged in parallel in separate ducts. The supply-air temperature is controlled by individually modulating hot- and cold-deck dampers to maintain the supply-air temperature. Like in the single-duct VAV, the supply-air temperature may be reset based on changes in building heating and cooling loads.

Packaged-Unit Systems

Prefabricated air-handling systems, commonly referred to as packaged units or rooftop units, have become commonplace in the building HVAC industry. These units are usually a single-zone variable volume system with a fan, a filter, heating- or cooling-water coils or direct-expansion refrigerant coils, and a damper arrangement for outside-air or outside-air and return-air mixing for economization. The problem with conventional controls on packaged units has been cost; in many cases, on smaller tonnage systems, the cost of field-installed controls has exceeded the cost of the unit. Packaged-unit manufacturers responded to this problem by providing basic control components on units at the factory for the control of temperature and pressure. The problem with these discrete devices was that they could not integrate with other

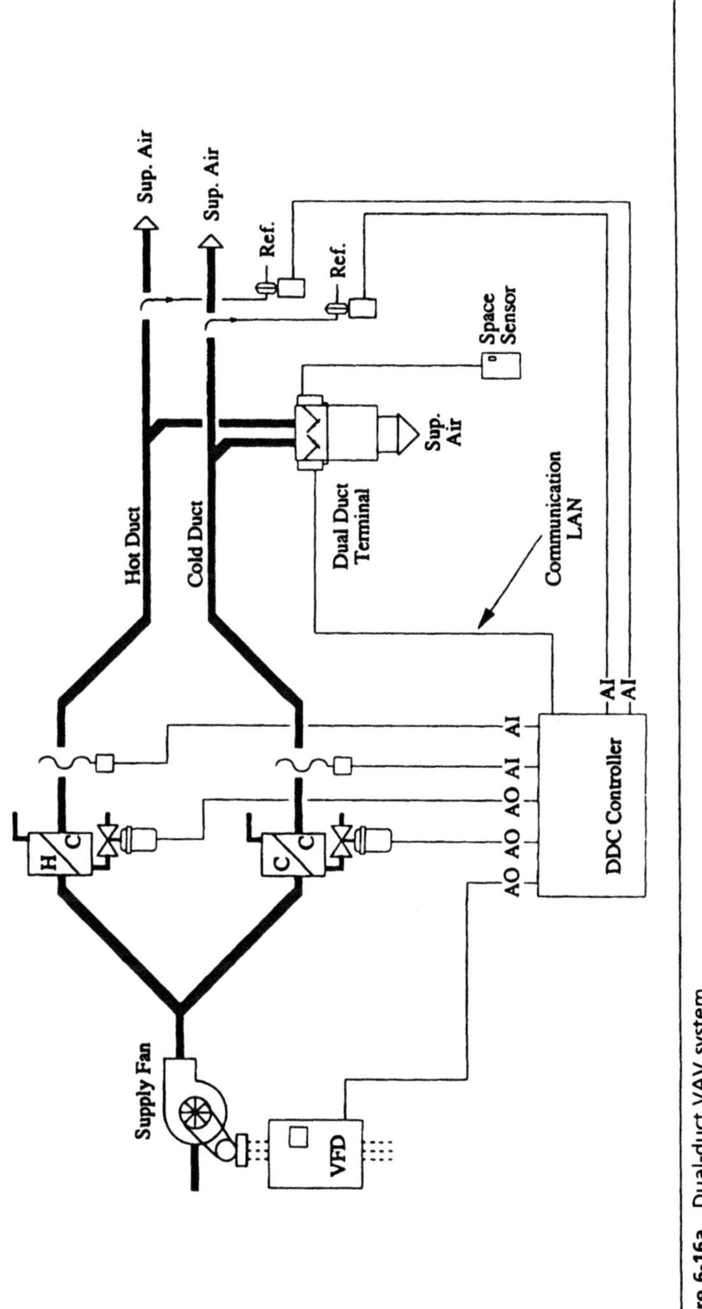

Figure 6-16a Dual-duct VAV system.

Point Description	(DO) Start/Stop	Open/Close	Other	Alarm	(AO) Position Control	Speed Control	Control Point Adjust	Other	Alarm	(DI) Status	Other	Alarm	(AI) Temperature	Pressure	Flow	Speed	Volume	Other	Alarm
Hot Duct Static Pressure (in.WG)														X					
Cold Duct Static Pressure (in.WG)														X					
Supply Air Temp., Hot Duct (°F)													X						
Supply Air Temp., Cold Duct (°F)													X						
Supply Fan Speed Control						X													
HW Coil Valve Position Control					X														
CHW Coil Valve Position Control					X														

Figure 6-16b Input/output checksheet: dual-duct VAV system.

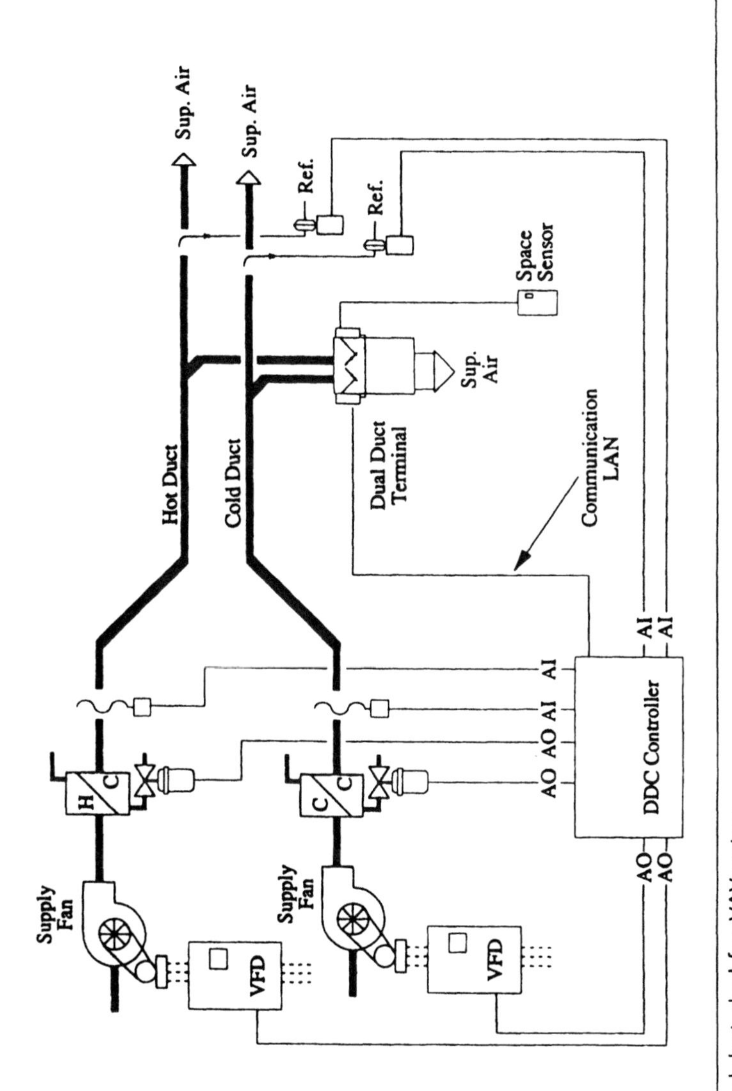

Figure 6-17 Dual-duct, dual-fan VAV system.

control devices in the HVAC system owing to product or power source incompatibility. Inability of building operations personnel to monitor analog data such as supply-air temperature or duct pressure from a remote location prompted control manufacturers to develop digital unitary controllers to provide stand-alone digital control of small air-handling systems. Being intelligent devices, they can perform complex control strategies at the unit level without program instruction from a central computer. Furthermore, by being able to share information on the operation of each packaged air handler over a local area network, each air-handling system in a building can be remotely monitored and supervised. With all sensors and field devices installed and compartmentalized into a single air-handling unit, the cost to apply digital unitary controllers to small air-handling systems has become competitive with the cost of conventional control methods.

Monitoring Strategies for Building Management

DDC systems have databases containing valuable information on the operating characteristics of the systems they control. In many instances, the value of this information is overlooked by design professionals and building operations personnel. Following are some examples of the kind of information that is available and how it can be used to improve the efficiency and effectiveness of a DDC system.

Power Consumption Metering

Lighting loads and motor loads for HVAC equipment consume significant electric power; when electric power costs are passed on to building tenants, or if such costs are segmented for cost accounting purposes, electric power measurement is required.

Consumption metering allows a building operator to segment the consumption of electric energy by zone, building section, air-handling system, or on a building-by-building basis on multibuilding campuses. So long as a means for measuring power is provided to the digital controller, these calculations can be performed to provide management information.

Input signals to the digital controller can come from several sources. Watt transducers, current transducers, and pulse signals from watt-hour meters are devices used for power measurement. In addition, by totaling the run times of HVAC equipment and central plant equipment, we can derive electricity costs by multiplying consumption rates against electric rate schedules.

Trend Logs

A trend log is a sequential listing of the current values of points on the control system. It is used to verify proper operation of the system and is an invaluable troubleshooting tool. Any kind of information available within the control system database can be summarized, sorted, evaluated, and printed out as a permanent record. Furthermore, alarm summaries can be generated to indicate the time, location, and nature of an alarm to the building operator. Trend logs are perhaps the most effective management tool offered by microprocessor control systems because they allow evaluation of system operation, the effectiveness of maintenance programs, and the response procedures taken by building operations personnel to alarm conditions.

Operation Reports

An operation report is like a trend log except that it provides information on a single system loop or a single HVAC system. Individual systems can be studied by collecting and summarizing data on their performances from the database of the control system. Problems such as excessive run time can be easily detected, and alarm strategies can be set up to prevent future problems.

Maintenance Time Reminders

Highly specialized *maintenance time reminder* (MTR) software has been developed for use on modern DDC systems to organize, monitor, and acknowledge the activities associated with a comprehensive HVAC system maintenance program. Equipment operation status, total equipment run time, and the date of last service are examples of the kind of information collected and summarized by MTR software. Sophisticated software packages generate preventive maintenance work orders when maintenance is required. When maintenance services have been completed, maintenance technicians must acknowledge this before the software will allow the control system to resume operation. Monthly summaries of maintenance inspections and minor repairs can be generated as a tool for management in the measurement of maintenance program effectiveness.

Energy Management Reports

In an effort to curtail the rising cost of energy, many building owners and operations personnel are implementing comprehensive energy management programs. Historical data on electricity usage can be summarized and archived by the control system as a means to evaluate the operating efficiencies of HVAC systems. Energy use profiles can be developed for each system in a building and used to estimate the potential saving that may be achieved by implementing energy reduction strategies, such as demand limiting or load shedding. With the trend log capabilities of the control system, energy consumption can be measured by time of day, by season, or by special event and used to project future demand. Sophisticated back-end analysis of energy usage can be broken down into categories such as lighting, air fans, mechanical cooling apparatus, and mechanical heating apparatus, all of which are heavy consumers of electric energy.

An important usage of energy management reports is for cost justification for energy management system (EMS) investments. Public institutions, such as colleges or universities, frequently fund energy management retrofit programs with bond money that is raised solely for the purpose of implementing an energy reduction program. The cash flows to retire the bond debt are obtained from the energy saving that such EMS strategies yield. Therefore, to determine the payback on the investment and the ability to repay the bond from projected energy saving, engineers must develop energy management profiles before, during, and after the application of EMS control strategies. Finally, energy management reports can be used to evaluate the effectiveness of experimental control strategies implemented through the computer-based control system. For example, a new chiller optimization strategy may be implemented based on a theory developed by the building engineer. By implementing control strategy changes through software, the building engineer can quickly determine whether a new control strategy is viable.

Supervisory Control Strategies

DDC systems collect information in a database that it uses to make future decisions based on programmed relationships. Local-loop controllers, communicating over a LAN, access such archived information frequently in the execution of local control routines. Supervisory control is independent of the local-loop control or system control level of the system, but it has the capability to override local programmed control routines when necessary to achieve the overall objectives of the control system.

The objective of supervisory control is to adapt the mechanical system operation to the actual needs of the building environment. This is accomplished by continuously monitoring the activities of the local-loop control routines and making minor adjustments whenever necessary. Common examples of supervisory control corrections include adjustments to derivative rates on proportional plus integral plus derivative (PID) control loops, reset of supply-air temperatures in large air-handling systems, and emergency overrides of system operations for smoke containment.

The purpose of computer intelligence is not only to respond to problems but also to anticipate future problems based on historical operating data. The fields of adaptive control and artificial intelligence have developed from early supervisory control systems. The most effective forms of supervisory control involve time-scheduled control of equipment, run times, optimized starting and stopping of HVAC equipment, and selective shedding of noncritical loads during peak consumption periods. Owing to the hierarchical nature of distributed control systems and the infinite variations of control routines that can be achieved through computer-based control systems, virtually any supervisory control strategy can be imposed on an operating control system.

Time-Scheduled Control

The operation of HVAC systems based on building occupancy are accomplished by using time-scheduled control strategies. Of all of the control strategies presented here, time control is usually the strategy that produces the greatest energy saving.

Lighting and HVAC loads are the largest electric loads in a building, with lighting loads accounting for as much as 50% of the total energy consumed. By staggering the start times of lights and fans, in-rush currents can be reduced, resulting in a substantial energy cost saving. Time-scheduled control can be accomplished by using devices that range from simple mechanical time clocks to sophisticated direct digital controllers. Examples of common time-scheduled controls are fixed time-of-day schedules, time-of-week schedules based on daily building occupancy schedules, and 8-day schedules that incorporate holiday or special-event schedules into a 7-day program. Microprocessor-based time clocks offer the ability to perform other timed resets, such as summer/winter building temperature set point changeovers, preprogrammed vacation schedule temperature set point changes, and special-event temperature set point changes. Fully distributed DDC systems offer the additional benefits of remote program overrides from a host computer or a local monitor, as well as allowing local program changes to occur at each field control panel. Recent developments have been made in voice/data interface using DDC software that allows a human interface over a touch-tone telephone to monitor and change system operating parameters. One voice/data product currently available provides two-way communication capability; the control system, operating through a modem, can access a series of emergency telephone numbers and relate specific event information, which has been compiled from a vocabulary of words stored in the computer memory to a listener. The receiver of the call acknowledges receipt of an alarm by pressing designated numbers on a touch-tone telephone.

Optimal Start/Stop

Optimal start / stop strategies are software-driven routines that calculate the optimal time of day and operating conditions for initiating system operation. The most common example of start/stop control is for the determination of building morning warm-up by referencing historical morning temperatures over a defined period of time. For example, the computer-based control system can determine the optimal time to start the building HVAC system in the morning to bring the building up to its proper temperature by the time occupancy occurs. Although this strategy does not produce the radical saving of time schedule control, it does offer the benefit of not starting the building system too early, which wastes energy, or starting the building system too late, which causes discomfort to early occupants.

Duty Cycle Control

Duty cycling is a strategy that involves shutting down selected HVAC systems for a fraction of their normal run time to reduce the overall electricity demand of the total system. The concept is that an energy-consuming HVAC system may not need to operate at full capacity to satisfy the building load. For example, a fan system could be shut down for 30 minutes twice a day on a staggered schedule with other fan systems, yielding several hours of energy saving without a noticeable effect on the occupants of the building.

The negative aspect of duty cycling concerns the effect of wear and tear on mechanical equipment caused by excessive on/off cycling. Furthermore, if the duty cycling schedule is too aggressive, it will result in annoyance and complaints from building occupants. It is for these reasons that duty cycling is seldom used in commercial HVAC applications. It is much wiser to use a motor speed control device to reduce fan and pump motor speeds rather than turning them off. Variable-frequency drives are the most energy-efficient means of controlling motors because motor energy consumption varies as the cube of the flow rate. Electrically modulating the speed of an electric motor is usually more effective in saving energy than cycling when in-rush currents and frequency of start and stop times are considered.

Load Shedding

Load shedding is a strategy that is similar to duty cycling control except that the command to start or stop a piece of mechanical equipment is a function of the overall building electricity demand. When energy consumption rates are high, usually during periods of peak outdoor-air temperature or occupancy, selected electric motors on mechanical equipment can be randomly shut down for brief periods to reduce temporarily the overall demand for energy.

Typically, loads are shut down for 15 to 45 minutes and then resumed. This is usually not enough time for a radical temperature change to occur in a controlled space. The time intervals of equipment shutdowns can be archived by the DDC system to develop a trend analysis of the total building demand that can be shed without affecting occupant comfort. In many metropolitan areas, utility companies charge a premium for energy consumption during peak periods of the day. This premium, or demand charge, can be as much as half of the total annual energy cost of operating the building. A properly implemented load-shedding strategy anticipates peak demands and will selectively shut down nonessential HVAC loads for short periods of time to avoid exceeding the demand penalty threshold. A DDC system measuring total energy demand can calculate the potential for exceeding a predetermined consumption ceiling. Using historical operating data in its memory, an intelligent system can compare current consumption levels to past consumption based on time of day and time of year, and can shed loads before demand penalties are triggered.

To be effective, this strategy requires most of the power-consuming equipment in a building to be able to be controlled by the digital control system. This includes HVAC equipment, lights, and any other single piece of equipment that has the potential of being a large energy consumer.

Temperature Setback

Temperature setback, also called night setback or night depression, is used to reduce the temperature of a building during unoccupied times. The purpose of this strategy is twofold. In colder climates, temperature setback is used to prevent a building from overcooling during the evening hours by continuing to heat the building at a lower set point than the set point for the occupied hours of operation. Usually night setback is 10 to 15° lower than normal set point. The HVAC system can easily bring the building temperature up in the morning when occupancy begins.

In warmer climates, temperature setback is used as an energy saving strategy. Where daytime outdoor temperatures are high and evening temperatures significantly lower, such as in areas of higher elevation, it is common to use cool evening air to precool the building. Precooling will usually keep the building within the comfort zone during the morning hours with very little supplemental cooling required by the HVAC system. By precooling, substantial energy saving can be realized by reducing the overall run time of mechanical equipment during occupied hours when electricity rates are usually higher.

Optimization Control Routines

Special software has been developed for computer-based control systems that are specifically designed to optimize or improve the performance of special-use HVAC equipment such as chillers and boilers. There are two types of optimization methods used. The first method involves improving the efficiency of the equipment that is being controlled, and the second method is based on improving the system efficiency by matching the mechanical device to its load.

System efficiency programs are used to reset discharge temperatures from devices such as boilers and chillers based on feedback from the HVAC system to match the output of the mechanical system to the actual demand for heating and cooling. By monitoring building loads the control system can reduce the overall amount of energy consumed while satisfying the comfort needs of the building occupants. Because most mechanical equipment, such as chillers and boilers, is manufactured with internal control systems, whose purpose it is to improve the efficiency of the operation of the individual device, it is more common to see software programs written to improve the efficiency of the overall chilled-water and hot-water plant operation than it is to see programs written for specific devices.

This chapter discussed specific strategies for local-loop control and the application of multiple-loop strategies to the control of entire HVAC systems. It investigated monitoring strategies and supervisory control strategies that have been proven to reduce energy consumption. This chapter has served as a survey of popular control methods utilizing the power of microprocessor-based controllers.

Chapter 7

Designing Direct Digital Control Systems

This chapter defines a methodology for evaluating and selecting the attributes of a direct digital control (DDC) system for any building project and for determining HVAC control requirements. The use of microprocessor-based controllers in HVAC control applications has developed significantly over the past decade as a result of (1) the ever-decreasing first cost of such systems; (2) more experience on the part of consultants, equipment vendors, and installers; and (3) overcoming the initial resistance by the HVAC industry to this new technology by the growing number of successful DDC systems in operation. A properly designed control system improves HVAC system performance and provides the building owner and operator with a valuable energy management and facilities management tool.

The objective of such an evaluation is to provide the building owner with a fully functioning, cost-effective HVAC system that meets the present and future needs of the building occupants. Furthermore, a properly designed and specified DDC system will lower the first cost by attracting competitive bids from qualified vendors. The criteria for design considerations described in the following paragraphs are offered as a framework for decisions involving DDC systems and their application to mechanical systems.

Criteria for Evaluating System Needs

Control system selection begins and ends with the fundamental control problems of a project. The design of a control system should not be attempted until a thorough knowledge of the building HVAC system has been acquired. Evaluating each piece of mechanical equipment in the system allows opportunities for improved performance control and energy savings to be easily recognized. The field of management science has provided us with a simple five-step approach to problem solution:

1. Define project objectives.
2. Define control system problems.
3. Develop alternative methods of problem solution.
4. Evaluate alternative methods of problem solution.
5. Select the best method.

Define Project Objectives

The objectives of an integrated control system project include cost-effective control for building comfort, expansion or modernization of an existing facility requiring updated and energy-efficient control technology, or improving the comfort and ability to control a large facility with a history of high energy cost. Other objectives may include applications for high reliability and extreme accuracy in control operations such as a clean room or surgical theater. Whatever the characteristics of the project in question, a list of primary objectives must be outlined at the outset of the design process.

Define Control System Problems

At this stage of design, a broad-based outline of control sequences should be applied to all subsystems in the HVAC system. The control of air volume, duct pressure, supply-air temperature and humidity, and sequences for smoke control and evacuation are primary considerations. Secondary control strategies for energy management, supervisory control and monitoring, override controls for building occupants, and other such strategies should also be thought out at this stage. The question should be asked: What is the minimum level of control required to render the system fully operable and to meet the needs of the building owner and occupant? The answer to this question forms the design intent of the project and allows for the definition of alternative methods of solving the stated control problems.

Develop Alternative Methods of Problem Solution

Once the basic control strategies have been identified, the evaluation of the appropriate method of achieving such control is more of an economic and functional process rather than a pure design decision. The decision must be made as to whether a centralized distributed or local control system is required. Furthermore, pneumatic, electronic, and direct digital technologies must be evaluated to find the most appropriate mix for achieving the control strategies. The designer should make rough sketches of the HVAC systems so that a comparative evaluation of all control technologies can be made.

Evaluate Alternative Methods of Problem Solution

Once alternative methods have been identified that can adequately solve the stated control problems, the evaluation requires a pragmatic approach. First-cost considerations are, of course, the most important in the minds of consultants, contractors, and building owners. However, other issues must be taken into account as well. Considerations must be given to the expected useful life of the building and to the type of tenancy. Long-range thinking is required to allow for future expansion needs by the building owner or potential new owners of the building. The accuracy of control operation must also be evaluated. Experience has shown that the best control system solutions utilize a mix of control technologies and control strategies. Each control system technology must be used where it is most cost-effective and efficient. For instance, with the familiar sensor/controller/controlled device format, an electronic sensor, an electronic controller, and a pneumatic controlled device acting through a transducer may be more cost-effective than an all-electronic system. An example of this is pneumatic actuation of large dampers controlled by a DDC controller based on an input from an electronic sensor. This application takes advantage of the cost effectiveness of pneumatic control devices and the accuracy and repeatability of a direct digital controller.

Select the Best Method

Once a control design strategy has been selected based on performance and economic factors, the process of detailed control system design can begin. However, a word of caution is in order: be prepared to reevaluate control strategies based on information that conflicts with the basic assumptions made when the design intent was drafted. Environmental factors affecting the project can alter the choice of a given control strategy. For example, in an existing facility, the availability of compressed air for pneumatic controls may be the basis of the decision to use a combination of pneumatic and digital controls. If it is discovered that the source of compressed air is insufficient to support the power needs of the control system, a totally digital system may be the best alternative. The design engineer should carefully research the characteristics of a project not only at the outset but also throughout the design development and final design stages.

Control System Design Considerations

By evaluating the proposed occupancy of the building, the designer can develop a "wish list" of monitoring and control functions. In new construction design, calculations of proposed energy consumption rates will identify opportunities for energy-efficient application of controls. For existing buildings, a complete review of the operating condition of the mechanical equipment and historical energy usage is required. This evaluation will uncover areas requiring improvement and will provide a platform against which future performance can be compared.

Regardless of whether the control application is new construction or replacement of an existing system, consideration must be given to the building operations staff who will use the control system on a day-to-day basis. The experience, training, and degree of sophistication of the building operators should be reflected in the design of the control system. A system that is beyond the comprehension of those who will use it serves no purpose; furthermore, experience shows that complicated systems are usually mechanically bypassed shortly after their warranty period expires!

Building operators charged with using a microprocessor-based control system should possess a basic understanding of computer fundamentals. To this end, a review of computer system fundamentals should be included as part of the operator training provided after the commissioning of the control system. The more a building operator knows about the system, the lower the likelihood of sabotage through a lack of understanding.

Evaluating Control Loops

The next step in the control design process is to evaluate the nature of each input and output of the control system. When all points in the system have been identified, they should be grouped into loops. Figure 7-1 presents a simple form used to identify and describe the point functions of a system. A complete list of the loops in each subsystem, as well as the overall system, serve as a reminder and checklist for the design engineer. A review of the control strategies detailed in Chapter 6 is recommended to make sure that every opportunity for performance and energy conservation efficiency is considered. The se-

System: AH-1

Loop	Description	Point Type
Loop #1	Supply Air Static Pressure	Analog In
Loop #1	Supply Fan Speed Control Output	Analog Out
Loop #2	Ventilation Duct OSA Volume Indication	Analog In
Loop #2	Ventilation Duct OSA Damper Control Output	Analog Out
Loop #3	Supply Air Temperature	Analog In
Loop #3	Outside Air Temperature	Analog In
Loop #3	Return Air Temperature	Analog In
Loop #3	Economizer Damper Position Control Output	Analog Out
Loop #3	Cooling Coil Valve Position Control Output	Analog Out
Loop #4	Return Air Volume Indication	Analog In
Loop #4	Supply Air Volume Indication	Analog In
Loop #4	Return Fan Speed Control Output	Analog Out

Figure 7-1 Sample loop assignment checksheet.

lection of control-loop strategies brings us to the next step in our evaluation—the determination of monitoring and control functions.

The monitoring and control functions of the system must be defined based on the value of the information the system can provide. The three most common monitoring functions are equipment status, energy consumption, and equipment run time. Other strategies that can be applied to loop control include supply-air temperature reset, based on information from occupied spaces, and night setback of building temperature to prevent overcooling from occurring during unoccupied hours. There are six basic groups of monitoring and control system functions:

- Data acquisition
- Timed control
- Feedback control
- Optimization
- Operator interface
- Off-line functions

Data Acquisition

Data acquisition functions provide measurement of controlled variables and information on the status of operating equipment, and safety and maintenance alarms; they report on climate conditions, energy usage, and trend control points in the system as directed by the software resident in the local control panel or host computer.

Timed Control

Simply stated, timed control means "turn it on; turn it off." Timed control of HVAC equipment is the simplest as well as the most effective means of reducing energy consumption in an HVAC system. Strategies based on timed control include demand limiting and load shedding.

Feedback Control

Feedback control is classic automatic temperature control using closed-loop control configurations. Control of supply-air temperature, mixed-air temperature, and duct pressure are all examples of closed-loop control. These control strategies and others were discussed and illustrated in Chapter 6.

Optimization

Optimization routines such as optimized start/stop, outdoor-air economizing for free cooling, and hot- and chilled-water plant optimization are all functions that can be performed by a DDC system.

The current trend in the manufacture of HVAC equipment, such as chillers and air handlers, is to provide packaged optimization controls from the factory. The advantage to this is that factory labor costs are much lower than field labor rates, and there is one source of warranty responsibility. The disadvantage is that packaged control panels often cannot communicate operating information to the building control system. To monitor the loading of such equipment, one may need to install redundant sensors. The advantage to field installation of controls is that all devices in the control system are of similar quality and are provided by a contractor with sole responsibility for the operation of the control system. There is no "finger pointing" between parties if a control-related problem emerges. Furthermore, all points in the control system can be accessed by the DDC system for control, monitoring, and reporting sequences.

Whether to integrate fully such control routines into the DDC system is based on how important such information is to the building operator in the overall scheme of building monitoring and control.

Operator Interface

Decisions have to be made concerning the level of operator interface desired and the tools required by the building operators to perform supervisory and control functions. At the very least, an operator's console or monitor must be provided. Remote telephone access to the system by means of a modem is also desirable, because this allows for remote assistance for system adjustments and programming. Graphic annunciation of alarms and maintenance reminders is also a recommended feature. Software that can report and acknowledge alarms, as well as upload and download data to and from the system, is also desirable and should be considered.

Off-Line Functions

Off-line functions refer to things that the host system can do when the HVAC control system is not in operation. The most important off-line functions are the collection and organization of operating data from the system, which can be in the form of hard-copy reports or data archives in the memory of the machine. This historical information can give building operators and engineers a look at the efficiency of their systems over time. Other off-line functions include the generation of graphic software and graphic screens as well as other custom programming that may be required to yield the greatest efficiency from the DDC system.

Evaluating Design Alternatives

After alternatives have been identified that meet the objectives of the project, each must be evaluated on the basis of economics, practicability, code requirements, requirements of operating personnel, and, in the case of an existing facility, the effects that construction will have on the interruption of building services.

The economics of the system must be favorable in terms of the energy and labor dollars saved versus the equipment and installation cost of the system. Chapter 9 provides guidance for estimating the installed cost of a complete DDC system. The other considerations mentioned fall into the category of common sense and should be incorporated into one's thinking whenever installation of controls is being considered.

The potential for future expansions to the system must again be carefully considered. As has been mentioned throughout this text, one of the most frustrating problems associated with the use of microcomputer-based control systems is the inability or the great expense required to add points to the system in the future. It is recommended that spare control points be provided on the system when it is first installed. The quantity of spare points is a function of the anticipated future expansion of the building, and should be decided by the design engineer and the building operator during the system design stage; as a rule of thumb, a minimum of 15% of each point type is recommended to accommodate any unforeseen point additions. When future expansions to the system occur, such modifications are simply a matter of adding field devices, connecting them to the DDC panel, and programming the DDC panel to accept these new points in its control routines.

Another important consideration is the availability of skills needed to operate a computer-based control system. The costs of hiring and training building operations personnel must be included in any economic calculations made at this point. In existing facilities, the operations and maintenance personnel charged with running the system must be trained not only in how the system operates but also on the fundamentals that the system is based on. If the building operations personnel are not fully supportive of the DDC system, then the system will eventually fail because of misapplication, misunderstanding of control strategies, sabotage, and neglect.

System Design Methodology

The design of conventional control systems, in general, is accomplished by completing the following steps:

1. Create HVAC system schematics.
2. Identify sensors and transmitters.
3. Identify controllers.
4. Identify controlled devices.
5. Identify the appropriate interconnecting media.
6. Specify sequences of control operation.

With DDC systems, the selection of control equipment is affected by the type, function, and location of control points in the system. The simplest way to organize the devices, sequences, and hardware requirements of a large control system is to use the point chart in Figure 7-2. To facilitate the control system design process and to avoid costly omissions, we suggest the following eight-step method:

1. Create the HVAC system schematics.
2. Identify each control point.
3. Identify sensors and transmitters.
4. Identify controlled devices.
5. Identify the appropriate interconnecting media.
6. Determine the sequence of control operation.
7. Identify the field computer hardware required to accomplish the specified sequences of control operation.
8. Identify the host computer hardware required to provide supervisory and monitoring control of the field control system.

Point Description	(DO)					(AO)						(DI)				(AI)						
	Start/Stop	Open/Close	Other	Alarm		Position Control	Speed Control	Control Point Adjust	Other	Alarm		Status	Other	Alarm		Temperature	Pressure	Flow	Speed	Volume	Other	Alarm

Figure 7-2 DDC system input/output checksheet.

Create HVAC System Schematics

The first step in the design process is to identify each component within the proposed or existing mechanical system, know how it operates, and know how it should be controlled. The simple air-handling system in Figure 7-3 illustrates the arrangement of HVAC equipment. Similar schematics should be developed for each subsystem in an HVAC system so that the design engineer can observe the arrangements of all equipment in a system at the same time.

Identify Each Control Point

A 'control point' denotes each field input or output to the digital control system. Control points can be physical points or software points (also called pseudopoints). A physical point is an actual input or output device that is physically wired to the digital control system through a terminal strip. A pseudopoint is a value or calculation product that is recognized by the digital controller. An example of a pseudopoint is the position of a damper actuator. The damper actuator is a controlled device that is assigned a physical output signal by the digital controller; the determination of the position of the actuator is accomplished by comparing the current value of the control signal to the known stroke of the actuator. It is important to understand the differences between physical points and pseudopoints when one is evaluating the point distributions of DDC systems.

From the point chart in Figure 7-2, each point in a system can be organized by loop assignment and by control function. For example, the static-pressure control loop requires an input from a static-pressure tip and an output to control fan speed. By logically grouping loops by function, an engineer can easily evaluate the modes of control and the degree of sophistication of a system.

Identify Sensors and Transmitters

The next step is to determine the type and location of sensing needed for each control point in the HVAC system. The location of sensors in duct systems or occupied spaces is a critical determinant of how well a control system will perform and determines how the system is to be interconnected. A discussion of sensor location with respect to system performance is provided in Chapter 2.

Identify Controlled Devices

Identify and locate controlled devices in the HVAC system. Again, use the point chart to summarize the arrangement of these devices to avoid omissions.

Identify the Appropriate Interconnecting Media

Give careful consideration to the selection of interconnecting media. Local building codes and the National Electrical Code provide guidelines for the design engineer and should be checked. Often, the most economical means of wiring or tubing a control system are not acceptable. For example, 18-gauge, three-conductor cable is used to connect sensors to local control panels; however, it may not be installed in mechanical equipment areas or in ceiling spaces that are not accessible. Care must also be given to sizing the conductors properly for their intended service. A methodology for selecting and sizing interconnecting media is discussed in Chapter 4.

Determine the Sequences of Control Operations

From the diagram in Figure 7-3, individual sequences of control can be developed that will eventually represent the entire control sequence for the project. By following these written sequences, we can break down each control mode within the system into logical loops. For example, supply-air control, mixed-air control, and static-pressure control are all examples of loops described within control sequences. The sequence in Figure 7-4 is an example of how the design engineer should describe control system operations.

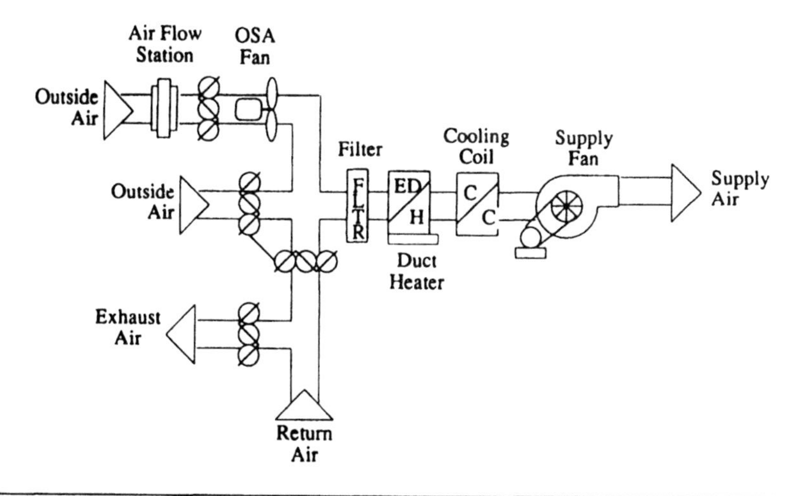

Figure 7-3 Mechanical system schematic example.

Operation Control:
The DDC Controller shall operate the supply fan system on a regular 365 day per year basis for both the regular and morning warmup modes of operation. The ventilation duct fan shall run when the supply fan system is running in the regular mode of operation. A smoke detector mounted in the return air stream just upstream of the exhaust air outlet shall be electrically interlocked to stop operation of both the supply fan and the ventilation duct fan and close both the ventilation duct damper and the outside air damper upon detection of smoke.

Ventilation Duct OSA Volume Control:
When in the regular mode of operation, the DDC Controller shall position the ventilation duct outside air damper to maintain the programmed minimum quantity of outside air to the mixed air plenum regardless of supply fan speed. The ventilation duct air flow station shall provide input to the DDC Controller for indication of outside air volume.

Supply Air Temperature Control:
When in the regular mode of operation, the DDC Controller shall position the economizer dampers to admit increasing amounts of outside air to the mixed air plenum (and decreasing amounts of return air) when the outside air temperature is below the return air temperature as a 'first' call for cooling as determined by a supply air temperature sensor. When the economizer dampers are in the full outside air position, the cooling coil control valve shall be positioned to allow increasing amounts of chilled water to flow through the coil to satisfy any demand for cooling When the outside air temperature is above the return air temperature the economizer dampers shall be placed in the full return air position and the cooling coil control valve shall be positioned to allow increasing amounts of chilled water to flow through the coil to satisfy any demand for cooling.
The setpoint for the supply air temperature shall be reset up or down based on the demand for cooling communicated to the DDC Controller through the Local Area Network (LAN).

Figure 7-4 Sequence of operation.

Identify Field Computer Hardware

When point charts for each HVAC system in a building system are combined, the architecture of a DDC system begins to take shape. The point density of each field panel must be carefully evaluated to make sure that the best grouping of points on a system is accomplished. Points within a single loop or related loop strategies should be kept in the same field panel. This ensures that, if a remote field panel fails, the execution of loops controlled by other field panels will not be affected. Another reason for keeping related loops on a single field panel is to keep system hysteresis from slowing down the response of a control loop. A heavily loaded local area network carrying communication signals between remote field panels may cause problems in fast-response control loops because information required by the loop to perform its control sequence may not be available as fast as is needed to perform such control. When the density and arrangement of field control panels are considered, the architecture of the system takes shape. Figure 7-5 presents a completed control layout for a single air-

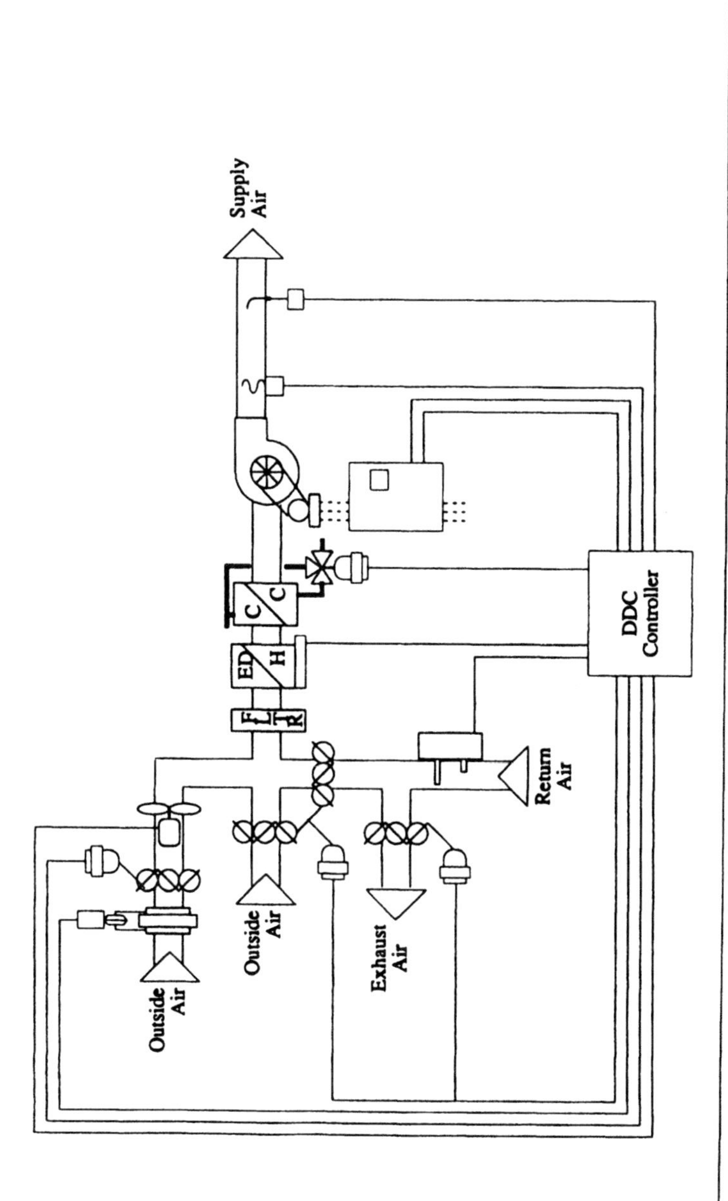

Figure 7-5 Control layout example.

handling system as a model. Similar layouts should be developed for each subsystem in the HVAC system.

Identify Host Computer Hardware

The last step is to define the supervisory and monitoring functions that the building owners and operators require of the system. Functions such as maintenance time reminders, alarm reporting, trend reports of energy consumption, tenant subbilling, and optimal building start/stop are examples of supervisory functions performed at the host level. The complexity of the desired functions for a system also defines the parameters of the software required to perform such control.

The design of temperature control systems should be done with a bottom-up approach. That is, start with the primary elements of the HVAC system and apply control modes to the system in a layered fashion until all control sequences required of the project have been incorporated into the design. The elements of a complete control design include detailed schematic diagrams indicating the arrangement of all HVAC equipment, control sensors, and control devices; complete written sequences for each mode of control describing in detail each control loop within subsystems; and a point identification chart indicating the function, type, and control action of each point on the system. Furthermore, it is recommended that a schematic of typical field control panels be provided on the contract drawings to indicate acceptable arrangements of points on field panels and to define the maximum point density allowed.

Chapter 8

Specifying Direct Digital Control Systems

Regardless of whether you are a consultant, a contractor, or a building owner attempting to upgrade or replace an outmoded control system, it is absolutely necessary to develop a specification to describe the system you want and how it should perform. There are few standard specifications available for computer-based control systems. Many that are available from industry and professional sources have been developed from negative past experiences with products and their installing contractors and are very conservative and quite defensive.

The most visible difficulty of using computer-based control systems concerns the lack of product standardization in the industry. It is important to understand why control manufacturers design their systems with hardware and software that is incompatible with competing systems in their industry. This strategy, known as *product differentiation*, reduces competition and increases profits to offset the high costs of product development.

With new milestones being passed in the design of microprocessors every year, control manufacturers are finding the average life cycle of a direct digital control (DDC) product line to be shorter than ever. This creates a support problem for users of outdated or obsolete systems that typically are no older than 2 to 3 years. Although these existing systems continue to operate and fulfill their purpose, they are technically out of date and do not receive the attention from the manufacturer that newer generation products command.

HVAC professionals must use extreme care in the selection of a system to avoid problems with system reliability, software, manufacturer support, and owner understanding of the system. By using the guidelines given here as a decision framework, you can avoid many of the risks associated with the application of computerized control system.

There are three primary steps in the specification process for computer-based control systems:

1. Evaluate and select a system architecture.
2. Evaluate and select control manufacturers.
3. Prepare specification documents.

System Architecture and Product Evaluation

There are over one hundred manufacturers and assemblers of proprietary microprocessor-based control products in the United States. Although the primary purpose of this equipment is to control HVAC equipment and manage energy consumption, there are major differences between these devices in terms of quality, performance, flexibility, and intelligence. Complete information must be gathered by the specifying engineer when selecting a building automation product for an HVAC application to be sure that the product has the necessary features required to satisfy the needs of the control application. There are, unfortunately, many circumstances where an improper match takes place between the equipment and the application, and this misapplication of equipment happens at every level in the building design process. Mechanical design consultants, mechanical contractors, control contractors, and building owners have all suffered financial losses from the misapplication of control system components. Therefore, it is the purpose of this section to outline effective criteria for evaluating control applications and to advocate defensive specification techniques.

Distribution of System Intelligence

One of the most important considerations that can be made concerns the location of the intelligence in a DDC system. The way in which microprocessor-based controllers are distributed throughout the architecture of a computer-based control system has a lot to do with how well the system will work once it is installed. Distributed control systems offer the advantage of reducing the risk of system failure due to the failure of a single component in the system. The failure of a 16-point direct digital controller has much less of an impact on the overall performance of the HVAC system than, say, a 64-point direct digital controller that fails. Therefore, it is important to consider the density of point groups per microprocessor in a control system architecture. The fewer the number of points per direct digital controller the greater the control system is said to be "distributed." Not all DDC panels contain intelligence. *Data-gathering panels* (DGPs) or *data acquisition panels* (DAPs) are simple enclosures that house termination strips that connect to sensors and control devices. They facilitate communication from a central controller to the field devices being controlled. The intelligence to interpret this incoming and outgoing data is located in a central control panel. It is important to consider how many data-gathering panels are serving a single microprocessor because, in essence, this is the number of points that can be lost in the event of a microprocessor failure. Another important consideration concerning the distribution of microprocessors throughout a system architecture is the location of the programs that control the operation of each intelligent control panel. True direct digital controllers contain memory chips holding application programs that tell the direct digital controller how to perform specific control routines. The advantage of having program routines at the local level is that control sequences for HVAC equipment can be installed in a digital control panel dedicated to serving that equipment. With distributed programming, many systems can be controlled independently yet tied together through a local area network back to a supervisory or monitoring computer. Distributed programming also reduces the amount of time required for the computer control system to execute program routines because program commands are not downloaded from a central computer to the point of control; rather, they are sent from the direct digital controller directly to the devices under control (hence the term *direct digital control*). This reduces the amount of traffic on the local area network and substantially increases the response time of the controller to a change in a controlled variable.

System Point Configuration

The point configuration of the individual DDC panel is also an important concern. Consideration must be given to the type and quantity of points on each HVAC system and how these points will be best distributed throughout the DDC system architecture. It is important to utilize available input/output (I/O) points in a DDC panel in the most efficient manner. The reason is that most systems utilize separate microprocessor I/O termination cards for each type of point. Therefore, the minimum number of cards in a typical direct digital control panel is four:

One analog input card
One analog output card
One binary input card
One binary output card

If the quantity of any of the four kinds of points exceeds the capacity of the cards mounted in the DDC panel, additional cards will have to be added to accommodate these points. This increases the cost of the DDC system and can create problems with respect to the ability to add control points to the system in the future. One of the most serious frustrations suffered by building operators is the expense, and often the inability, to add additional points to an existing DDC system. When properly designed, a minimum quantity of spare points of each type should be specified to allow building operators the flexibility to add more points to the system in the future without requiring that additional hardware be added to the system.

Man–Machine Interface

Some intelligent field panels offer an integral display that allows system data to be read by a building engineer at the local level. Information on the operation of the entire system can be accessed at any intelligent field panel via the control keypad and display. This interface can also be in the form of a hand-held terminal that can be connected to the local control panel through an RS-232 communication port.

It is important to consider the format in which system data are presented at the local level. Obviously, the more the data assimilate the English language, the easier they will be to understand. Common display formats range from hexadecimal information, which is very difficult to interpret in the field, to "friendly" systems that use full English descriptors with commonly used engineering terms.

The degree to which localized information is accessible to the user is related to the word length limitations of the digital controller. Variable names more than 10 characters in length need abbreviations, which means that the building operator must memorize or catalog common abbreviations to review information on the display. Variable names such as "chilled-water supply temperature" can be abbreviated to read "CHWS TEMP," but "east wing lobby temperature" requires more characters.

The format of local information should be described and illustrated in the product literature of the control system manufacturer, and it should be checked to make sure that building operators will be capable of understanding and utilizing such field data.

Acceptable Manufacturers

Once the architecture of the control system has been established and the primary screening of manufacturers has taken place to determine which control systems are best suited to the HVAC application in question, the next step is to develop a list of acceptable manufacturers who will be allowed to bid. Obviously, many control companies will want to bid a project, each feeling that its system has the capability to perform to the standards specified. Unfortunately, this is not always the case from the standpoint of the building operator. As difficult as it is, choices must be made as to which companies will be allowed to bid. Important considerations that go into this decision process are:

- The quality standards to which products and systems are manufactured (ISO 9000, 6 SIGMA, etc.)
- The number of projects the manufacturer has installed that are similar in scope and architecture to the HVAC system in question
- The number of years the manufacturer has been in business manufacturing the same type of equipment that it is now offering for sale
- The support network, at the factory level and the local level, available to the installing contractor and the building owner after installation
- The availability and location of replacement parts for the system installed
- The experience and reputation of the installing control contractor

The most reliable source of information on the quality and performance of a brand of control product is the existing customer base of that manufacturer. Although a dissatisfied building owner may not always be the fault of the control system manufacturer, it is an indication of the level of acceptance for a brand of product and is a statement about the commitment that the manufacturer makes in ensuring that its products are properly applied and installed. It is wise to check the references of the manufacturer by contracting existing users of the product that have systems similar to the one in question. Most manufacturers are more than willing to provide a reference list to a prospective customer in hopes that the positive response received from existing customers will encourage a decision to purchase their system. Because manufacturers want to put their best foot forward, the reference list will include only those projects that are known successes. Therefore, it is wise to contact mechanical contractors in the local area to query them about their experience with a particular brand of controls and the controls contractor who installs them. Mechanical contractors have much more exposure to control system contractors than most consultants and building operators, and they can provide candid background information that is unavailable from project reference lists provided by prospective manufacturers. Taking a tour of existing control installations for each prospective bidder is also recommended. In many cases, not enough time is available to perform such extensive research and evaluation prior to bid. In such circumstances, a mandatory walk-through of two or three existing projects of the apparent low bidder by the owner's consultant prior to award of a contract is recommended. Failure to demonstrate experience and competence in the installation of systems similar to the system under consideration is a reason for the consultant to reject the low bid.

Specifications

To ensure that the control system that gets installed on your job will perform to the minimum standards specified, you should write a strongly worded specification that spells out the requirements of the system in detail. Two primary types of specifications can be written for a DDC system: proprietary specifications and functional specifications.

A *proprietary specification* is written around a specific product or system after the product has been selected by careful review of all brands available for a job. It describes in detail the features and capabilities of the selected control system and sets the standard by which other products will be judged. The advantage of a proprietary specification is that it ensures the building owner that he will receive exactly what he wants. The disadvantage of this type of specification is that once the control product manufacturer realizes that he or she has a proprietary specification written for the product, he or she may charge an unfair price for the system.

Functional specifications, also called *performance specifications*, describe in detail the exact system features, performance levels, and minimum standards that the control system must have, but they do not include any specific references to a brand of product. Functional specifications can be used to eliminate unqualified bidders by requiring control performance that is beyond the capability of the substandard systems. Functional specifications are frequently used on public works projects where proprietary specifications are not allowed. The advantage of functional specifications is that they encourage open bidding. The disadvantage is that if the specifications are not written carefully, an unscrupulous contractor can take advantage of generalized descriptions in the specification to provide a system that does not meet the owner's needs, even worse, a platform to generate costly project change orders.

Experience has shown that the most effective means of writing specifications is to combine the positive elements of proprietary and functional specifications to produce specifications that aptly describe the minimum feature set of the control system as well as the minimum performance requirements expected of the system installed. What follows is a suggested outline for a specification that addresses each issue discussed thus far. General comments have been incorporated into each section to explain concepts and to suggest ideas for improving specification documents. This outline is offered as a checklist only and should be expanded and adapted to meet the needs of each project.

Suggested Specification for Direct Digital Control Systems

The following specification language is offered as a guide to the specifying engineer preparing a specification to determine the minimum acceptable materials and methods of furnishing, installing, and supporting a DDC system.

Scope

The microprocessor control systems (MCS) as specified and shown herein shall be fully integrated and installed as a complete package by a single contractor. The contractor shall furnish all materials, including all computer hardware and software, operator input/output peripherals, field hardware panels, sensors and field control devices, and installed wiring and piping. The contractor shall be responsible for engineering, supervision of installation, labor services, system calibration, initial software programming, and system checkout as necessary to provide a complete and fully operational MCS as specified herein.

Vendor Qualifications

The successful contractor shall have an office local to the job site staffed with factory-trained engineers and technicians fully capable of providing instruction, routine maintenance, and emergency maintenance service on all facets of the system. The contractor shall have a minimum 5-year experience record in the design, engineering, installation, and programming of computer-based building control systems. The contractor shall be prepared to provide a list of no fewer than 10 similar projects that have building management and control systems similar to the system specified herein. These projects shall be on-line and functional such that the owner's representative may observe a system in full operation that is the exact equivalent to the system proposed to be installed under this contract.

System Description

The control system shall be configured as follows.

System Structure

The MCS shall be modular in design consisting of centralized control equipment, operator peripheral devices, and field hardware panels interconnected via a communications network. The hardware supplied shall consist of physical components that are built to established industry standards, that support the communications scheme of this system, and do not require customization to meet the requirements of this specification. All hardware and software provided under this contract normally shall be held in the inventory of both the manufacturer and the installing contractor and shall be considered items of standard manufacture. Devices that are currently out of production or are slated to be phased out of production within a 24-month period from the date of this document shall not be considered for this project.

Operator Interface

The following paragraph outlines the requirements for the operator station, which is usually a microcomputer capable of interfacing over the network in a real-time, multiuser, multitasking format with the field control panels. Competing manufacturers use different operating systems as the environment within which their programs and software routines operate. The configuration of the specified host computer must match the requirements of the acceptable control systems. Consideration must be given to the integrity of the host computer selected as well as the graphics software available to be run on it. The host computer should be specified to include a minimum quantity of RAM memory, a minimum amount of hard disk storage space, minimum requirements for multistation and multitasking user operations, and a minimum clock-rate speed for the microprocessor upon which the computer is based. Modern graphics software packages usually require a minimum of 16 Mbytes of RAM memory, a minimum of 50 Mbytes of hard disk storage for program storage and trend data acquisition and collection, and clock speeds of 50 to 100 MHz.

Point Capability

The MCS shall be capable of supporting the type and number of hardware points listed on the summary of system points attached to the specification. In addition, the MCS shall support all software points necessary for the specified control functions.

This paragraph makes clear to the controls contractor that he or she is responsible to provide all data acquisition hardware required to accommodate the total point requirements of the MCS without additional cost to the owner.

Spare Point Capacity

The MCS shall be provided with a minimum quantity of 15% surplus of hardware points of each point type. The distribution of these surplus points shall be equal among all four types of points.

This paragraph prevents costly hardware additions resulting from insufficient hardware to accommodate the future point needs of the building systems. As stated earlier, this has been one of the key criticisms of DDC systems. Many times after the project has been installed, the building owner and operator find to their dismay that expensive additional hardware must be added to their system to accommodate even a few minor point additions. This paragraph will alert bidding contractors to the minimum requirements and expectations that the owner has for the future expansion of this system.

Expansion

Expansion of the MCS shall accommodate additional hardware components to meet future expansion requirements but shall not require special software to achieve this expansion that modifies the operation of the system or its operating systems.

Distributed Processing

Each field unit shall be capable of performing all specified control functions in a completely independent manner. If any other field unit or communications processor malfunctions within the system, all other field units shall continue to control, monitor, and have the ability to be accessed and programmed without degrading the overall performance of the control system.

This paragraph determines the maximum number of points per field panel which in turn defines the limit of the effect of a system failure. For example, if a distributed processing system is broken up into units of 16 points per intelligent field panel, the maximum effect from the loss of a single panel is 16 points off-line. As most building control systems comprise anywhere from 50 to 1000 points, the effects of a single panel failure in this example is of less importance because of fewer points off-line. However, some systems use a larger number of points per panel. This point distribution is an important selection criterion that must be evaluated carefully when selecting acceptable bidders.

System Networking

Each field control panel shall be capable of sharing information with other field control panels such that control sequences or control loops executed at one control unit may receive input signals from sensors connected to other units within the network. If the network communication link fails, or the originating field control panel malfunctions, the control loops shall continue to function using the last value received from the failed component. Again, this paragraph and the preceding one define the maximum impact on system performance in the event of a failure of a single field control panel.

Wiring Materials and Methods

All control wiring shall be in thinwall grounded metallic conduits and shall conform to all applicable codes and the requirements of Division 16 Electrical. Provide type THHN conductors of not less than 17-gauge stranded, with at least one spare two-wire circuit in each conduit run. Control interlock wiring greater than 100 volts shall be a minimum 14-gauge stranded conductor with 600-volt insulation. Adequate overcurrent protection shall be provided.

Control cable shall be allowed in spaces that are concealed yet accessible provided that conditions for such use conform to local codes.

Control power and signal wiring shall be segregated throughout the installation of this project, and shall be kept at least 3 feet apart at all times. Under no circumstances shall power and signal wiring be run in the same raceway, regardless of the type of wire insulation. Control signal wiring shall be routed clear of all lighting ballasts and other electromagnetic devices that may damage the integrity of the control signal.

Be specific about the wiring requirements for the project. Incorrect wiring materials or methods can affect system performance by causing slower local area network (LAN) speeds, interfering with control signals, and impairing the quality of information collected by the host computer. Refer to local building codes for restrictions on the gauge and type of wire that are available for an application.

System Response Time

The intent of this specification is to provide the operator with a system that allows him or her to view the status of all points in a dynamic fashion over a graphic display on the operator's terminal. The maximum response time for acquiring point data in creating a graphic display shall not be greater than 3 seconds.

This paragraph defines the maximum amount of time for a change in a controlled variable to be updated at the graphic display. This response time is called the *refresh rate.*

Warranty

The controls manufacturer and the installing contractor shall provide a warranty covering all parts and labor, following the commissioning of the control system, for a time period to be determined by the owner, based on the unit pricing for such warranty submitted with this bid. Any warranty expense during this time period shall be borne entirely by the manufacturer or the installing contractor and shall include travel costs and all living expenses associated with the performance of such warranty. The cost for service calls necessary to maintain the integrity of the warranty shall be included as part of the bid price. All warranty work shall be performed by factory-authorized service personnel.

One way to reduce the risk of a control system failing to meet the expectations of the owner is to require that the installing contractor and/or the manufacturer provide an extended warranty for the system. Most standard warranties are for 1 year from the date of acceptance of the project. By forcing installing contractors or manufacturers to extend this warranty period, the burden is upon them to make sure that they correctly apply control products and use proper installation methods to limit their warranty expenses and still be able to maintain a competitive price at bid time. It is in the owner's best interest to require a warranty extension as an additional alternative price on a bid to measure the value of adding this warranty to the standard warranty of the system. Besides, it is far less expensive to purchase additional warranty when the job is bid than it is to take prices to maintain and service the system when competition for the project has been eliminated by the fact that one manufacturer has been selected.

The way a design consultant specifies and evaluates extended warranty can have an impact on the value of the warranty services the owner receives. Extended warranty should be unit priced by year for the number of years the owner wishes to purchase. This can be done either as alternative additive pricing or as a deductive alternative on the base price proposal.

By fully defining and line-itemizing the requirements for extended warranty, all bidders are forced to recognize their obligation to provide this protection. All too often, when extended warranty requirements are made part of the base specification they are deliberately ignored by bidding contractors.

System Hardware

A general paragraph on acceptable system hardware should be included here. This should include the names of manufacturers who have been screened for acceptability and have been approved to bid the project. In addition to listing acceptable bidders, a note should be added that other bidders may be allowed to bid only if they have received prior approval from the engineering consultant or the owner's representative prior to bid time. It is not advisable to allow bids to be provided by unapproved contractors at bid time. This can create tremendous problems if a mechanical contractor should accept the bid of an unacceptable subcontractor for the control system portion of the project. There are many cases in construction law that involve the issue of manufacturer and contractor acceptability on public and private works. If an installing contractor proposes to install inferior material on a project and the owner or the consultant attempts to have this contractor removed, a significant cost differential could exist between the lowest bid and the next responsible bid for the project. This deficit is immediately passed from the subcontract level to the prime contractor and eventually to the owner, and can cause significant problems if the prime contractor or any of the subcontractors are forced to pull off of the job because of this price difference. If a bid bond was provided with the bids, the potential for litigation is increased because the owner will look to the bonding company to protect him or her from having to pay a higher price for the project. Therefore, it is crucial that all acceptable control contractors and manufacturers be spelled out in this portion of the specification as well as a procedure and a means for other bidders to receive an evaluation and a determination of acceptance or rejection before bids are accepted and reviewed.

Central Computer Station

The operator central computer station shall be based on an IBM, or equivalent microprocessor. The minimum acceptable configuration of a personal computer for this application shall be:

640-Kbyte RAM memory
Serial and parallel ports
1.2 Kilobyte floppy disk drive
2 giga-byte internal hard disk drive
Enhanced color display monitor
VGA graphics adapter with minimum 2 megabyte memory
Keyboard/mouse/operator interface to system
Internal modem; minimum 9600 baud for remote communication

As stated earlier, systems by competing manufacturers may require different hardware specifications at the host computer level. Therefore, each acceptable manufacturer should be consulted to make sure that the minimum requirements of each manufacturer are reflected in this list.

Computer Display Characteristics

If a central computer is desired for a control system project to assist the operator in interfacing and monitoring the control of his or her system, it is recommended that a color graphic display system be provided to allow graphic representation of real-time control situations. No tool has proven more useful for the average building operator than to have a picture of what is going on around him or her, to have the means through an interface device to acquire information on the status of a system, and to execute control commands from a console with the touch of a few buttons or keys. Therefore, appropriate language should be included to describe how the system should look, feel, and operate:

A dynamic color graphic display terminal shall be provided that shall display alphanumeric data and dynamic color diagrams simultaneously on one screen. The graphic display shall display real-time data, have the capability to generate color schematics of building equipment or areas being monitored, and allow operator commands and report system activity in a simultaneous fashion. The unit shall support a mouse or other operator interface device that will allow the operator to make menu selections, execute graphic commands, choose graphic symbols, and even make free-hand drawings of existing or new building systems.

The color graphic display terminal shall, upon command, generate color schematics of building equipment or groups of building equipment or areas being monitored and simultaneously display the current measured variables associated with the equipment or area. The color graphic displays shall be dynamic in that point data or calculated values will change continuously while under observation. Points in an alarm condition shall be identified by a distinct color or a flashing symbol that is easily identifiable by the system operator.

The computer displays shall be arranged in a format that allows easy user interaction. The display format shall be divided into separate areas, including a main menu, an alarm area, and the main display area.

Printers

It is important to specify the minimum requirements for printing reports and providing hard copy of other important data available from the system.

A narrow-carriage high-speed dot matrix printer shall be provided for hard-copy data and alarm printouts. The minimum acceptable print speed shall be 200 characters per second. The printer shall have a line length of at least 132 characters in a compressed mode and shall use a nine-pin head. The printer shall be of the read-only type.

Uninterruptable Power Supply

Uninterruptable power supply (UPS) is a source of electric power that operates in parallel with electric utility sources to supply an electric load without interruption when the utility source fails. A loss of power of even a few seconds can cause a catastrophic loss of data within computer-based control systems, especially if they are involved in routines that are collecting, monitoring, and manipulating field data in a real-time fashion. Most UPS systems comprise batteries operating an inverter connected to a critical load at all times. The batteries are being charged by a rectifier from the prime power source and therefore always maintain their charge. Where utmost reliability is required, two or more redundant systems of rectifiers and inverters may be connected in parallel. For computer-based control systems, it is recommended that a small UPS system intended for microcomputers be provided to allow the microcomputer to continue operating for a minimum of 5 minutes after power is interrupted. This should be more than sufficient to offset any power spikes, transients, or brief losses of power that may occur in a normal electric utility system and to maintain the integrity of the data stored in RAM memory of the system during the execution of control routines.

Communication Speed

The communication speed between the CPU and the remote field panels shall run at a minimum rate of 9600 baud or higher. It is important to determine the minimum speed at which data will transfer throughout a LAN or from field panel to field panel. This can have a great effect on the overall response rate of the microprocessor control system, especially where there are many remote field panels involved. Under normal circumstances, the greater the number of field panels and the more communication that takes place on a LAN, the slower the LAN will operate. Therefore, the minimum speed at which the LAN is capable of communicating is a critical determinant of how fast the system can refresh information.

Field Hardware Specifications

Individual field interface panels shall monitor and control all functions related to the DDC control system. Each panel shall be modular in design to allow for a phased installation and shall be designed using microprocessor components. Each unit shall have a pin-hinged door and master-keyed lock. Field units shall be capable of operating in an ambient environment of 32 to 100°F, 10% to 90% relative humidity.

The field interface panel shall be modular in design to accept multiple combinations of point types for monitoring and control of building automation functions. The control programs stored in the field interface panels and the mass storage unit associated with the CPU shall be properly protected. On systems that do not provide programs in firmware, a battery backup on both the disk memory and the field interface panels shall be provided for a minimum of 30 days.

This specifications in this paragraph prevent a loss of system control due to a memory failure or system error. Also recommended is a paragraph stating that:

Each field interface panel shall perform continuous diagnostics and any malfunction detected shall be announced at the operator's console as well as visually indicated at the field panel where the malfunction takes place. Transient surge protection shall be provided at each field interface panel for the purpose of suppressing voltage transients that may occur from natural or unnatural causes.

Information Displays

The actual values of all points on the system shall be readable through a digital display on the face of the field interface panel or at the CPU. All temperatures shall be displayed in degrees Fahrenheit. Other input and output signals may be displayed in either engineering units or the actual signal value. Where signal values are displayed, provide conversion instructions in equations to allow an operator to transfer such values manually into engineering units. Provide this information within the field interface panel enclosure.

The ultimate performance of a system rests on the accuracy and repeatability of its sensors. It is recommended that local temperature and pressure gauges be provided near the equipment being controlled to verify the accuracy of the data provided by the DDC system.

Sensing Inputs

Industry standard sensors with standards sensing signals shall be provided. Temperature, humidity, differential pressure, and other control signal inputs shall be one or more of the following types:

0 to 20 mA
4 to 20 mA
0 to 5 VDC
0 to 12 VDC
Variable-resistance signals (resistance temperature devices [RTDs])

The type of sensor selected for an application shall be considered with respect to calibration inaccuracy and sensor drift. In applications where abnormally long runs of wire are required to connect the sensor to the field interface panel, usually greater than 100 feet, a correction factor shall be provided within the field interface panel to offset any error caused by the increased resistance due to the length of the wire. In addition, any offset due to excessive ambient temperature and its effect on the conductivity of the sensor wire shall be factored into this correction formula.

Controlled Devices

Binary Outputs

Each field control panel shall have the ability to test the circuit driving the equipment contractor for the purpose of establishing whether or not an AC circuit to the contractor is active. Binary outputs, or relay outputs, shall be provided with pilot lights or light-emitting diode (LED) displays of the status of the signal.

Analog Outputs

Analog outputs shall be industry-standard 0 to 5 VDC, 0 to 12 VDC, 0 to 20 mA, or 4 to 20 mA.

Pneumatic Actuation

Pneumatic actuators may be used to control valves and dampers where the simplicity and reliability of pneumatic actuators may provide an advantage over electronically driven control devices. They may not, however, be used on equipment that normally accepts an electronic signal, such as the input signal to a variable-frequency drive. Positive positioners shall be provided on any pneumatic actuator controlling a valve larger than 1½ inch in diameter or a damper larger than 20 square feet to eliminate hysteresis caused by friction in the air lines or differential pressure in the actuated device. Gauges shall be provided on the face of any field control panel that is providing a pneumatic output and shall be appropriately identified with a permanent label.

General Programming Requirements

Programming Language

The DDC system programming language shall be a high-level language utilizing English-language commands and shall be designed for HVAC and energy management applications. The intent of this specification is that an operator with a working knowledge of such high-level languages as BASIC, Pascal, or FORTRAN, along with a general knowledge of EMCS control algorithms, shall be able to read and understand most of the program codes without having to reference outside documentation. Low-level, non-English languages that utilize encoded or encrypted codes shall be considered unacceptable.

This is an important requirement of the software for a DDC system. After the DDC system has been installed and is out of warranty, the building operator will be left the task of making program changes on the building control system. Therefore, it is important that the system selected be user friendly to allow the operator to perform minor program changes without relying on the manufacturer or a third-party computer programmer.

Standard Energy Management and Control Software

The DDC system shall be capable of executing the following standard control routines:

1. Time-of-day scheduling
2. Start/stop optimization
3. Demand-limiting routines
4. Temperature-compensated duty cycling
5. Economizer or free-cooling control
6. Enthalpy control
7. Supply-air reset
8. Chilled-water reset
9. Event-initiated programs
10. Local level trends

11. Maintenance time reminders
12. Failure alarms and exception alarms

These are suggested features and should be included or deleted to meet the control, monitoring, and reporting requirements of each project.

Custom Programming

The DDC system shall support the creation and modification of customized direct control algorithms based on arithmetic boolean, or time-based logic. Each unit shall support the creation, modification, or removal of control algorithms while the system is in operation. Each control loop shall be fully user-definable. Control algorithms shall be capable of being developed at a central PC for downloading to local control panels or may be programmed directly into individual local control panels.

Operator's Station

The controls contractor shall provide a software package to control and monitor the operation of the DDC system which is based upon a personal computer compatible with MS-DOS, which will provide the following functions as a minimum:

1. Graphic construction software
2. Point editing software
3. Program editing software
4. Window software
5. Help menus
6. Trend logs
7. Automatic upload and download of data to field control panels

Graphic Construction Software

The graphic construction software shall allow a building operator or system programmer to create customized graphic screens of building systems and control routines, allowing the selection and display of dynamic point values and the ability to locate them wherever desired on the graphic. The user interface to the graphic software shall employ a mouse to select and shape standard forms such as rectangles, circles, or lines. The software shall also provide standard graphic symbols to represent fans, air-handling units, or other industry standard symbols used to depict HVAC equipment that can be selected and modified as needed to meet the needs of the graphic being constructed.

Point Editing Software

The software system shall be provided with a full-screen point editor that shall allow the addition, deletion, and change of any point on the system. Complete point definitions shall be provided by filling in information in predefined information blocks within a single screen form. Improper or erroneous data entries shall be automatically disallowed.

Program Editing Software

Program editing shall be a full-screen, word-processing-type program designed to allow the operator to create or modify control sequences without having to reprogram the software completely.

Window Software

An appropriate software package for the host computer being provided shall allow the concurrent viewing and command of system operations under multiple screens or windows.

Help Menus

Help menus shall be accessible from all levels of the software package and shall be designed to provide help that is pertinent to the active screen.

This paragraph provides the system operator with active on-line help menus that shall provide enough information to preclude the need for detailed system documentation to provide answers to simple operating problems.

Trend Logs

System trend logs shall be capable of being viewed in an interactive mode and downloaded from the DDC system into a file format that is readable by an electronic spreadsheet, such as Lotus 1–2–3. File exchange formats such as SYLK or DIF are acceptable so long as they provide the complete transfer of alphanumeric characters to the electronic spreadsheet.

Upload and Download to Field Panels

Changes made to program algorithms into the DDC database that are made at the host computer shall be readily transferable to the individual field control panels that are affected by such changes. The collection of trend log data from the field control panels vertically through the DDC system architecture shall be fully automatic and shall not require individual data acquisition from each panel on the local area network.

Control Sequences

To be sure of obtaining a system that fully meets the operation expectations of the consultant, contractor, and building owner, it is absolutely necessary to provide a detailed sequence of control system operation. The control sequences should be described building by building, system by system, and unit by unit. For example, the control strategies for a complete building such as building lighting control, HVAC system performance optimization, and control of central plant equipment should be described in detail. Each system constituting the building HVAC system should be described individually. For instance, each air-handling system should be described, including such control strategies as economizer control, supply-air reset control, and fan speed control for VAV systems. Furthermore, descriptions of the individual control sequences to be performed at the zone level, such as the operation of a VAV terminal, are important and should be described in detail as well as supported with diagrams and control action charts on the mechanical blueprints. The written word usually takes precedence over any diagrams. The diagram simply communicates the fundamental nature of the system to a bidding contractor or installing contractor; however, the intent of the specification has not been met until the system can adequately perform the control sequences specified in this section. Therefore, it is imperative that a detailed control sequence be provided and frequently referenced where appropriate in DDC system specifications.

Enforcing Control Specifications

There has been considerable controversy in the HVAC industry concerning the way control specifications are written and the way that drawings are prepared to limit competition for control projects to only those vendors and contractors that are qualified to provide the system required. As a general rule, mechanical consulting engineers usually carry the responsibility for establishing a platform of quality that all bidders on the project must meet. Furthermore, the mechanical consultant bears the responsibility for evaluating and accepting or rejecting bids from control system vendors. Mechanical consultants act in the best interest of their clients by being extremely conservative in the selection and acceptance of control system products as well as in the evaluation of installing contractors. It is common to find that control system specifications written for DDC systems are very restrictive in their minimum requirement levels and, in some cases, quite proprietary. It is within the purview of the mechanical engineer's rights and responsibilities to decide what is acceptable and what is not on a project for his or her client. The mechanical consultant, however, must be careful to draw a fine line between product selection and product discrimination. State business codes and discrimination laws differ in the amount of authority mechanical engineers are allowed to exercise as they determine who is an "acceptable manufacturer." Usually, the engineer's rejection of inferior materials is supported by law, but discrimination against installing contractors is not. That is, a specific brand of product may be specified by name, but an installing contractor who is duly licensed and experienced in performing the type of work specified cannot be prevented from bidding on a project.

Project Funding

The source of funding for a project sometimes places constraints on the design consultant. Project funding comes from two primary sources: public funding and private funding.

Public projects, which are backed by public funds, are subject to regulations concerning discrimination against bidding contractors, whereas deliberately limiting competition between acceptable bidders is left to the discretion of the specifying engineer. Specifications for public projects are usually generic in nature and, although they may list acceptable manufacturers by name, are usually written in such a fashion as to allow any professional manufacturer of control systems or any professional installing contractor to bid on an equal footing with the specified system. Publicly funded projects present a special challenge to the design consultant as well as the building operator in that both parties, through their experience in the HVAC industry, may have a stated preference for one manufacturer over another. Although rules and regulations prevent outward discrimination, specifications can be written that satisfy the regulations of the public agency while limiting competition on the project to only those vendors who are truly qualified to provide the specified system. The most appropriate way to screen acceptable bidders is to present strict qualification standards as well as a mandatory job walk before bid date. Qualification standards such as minimum experience requirements for projects similar to the one in question, as well as the requirement to submit extensive qualification statements, are ways to make sure that control contractors interested in bidding the project are capable of performing to the satisfaction of the consultants and building operators. The requirement for a mandatory job walk prior to bid is an excellent way to perform a face-to-face interview with prospective bidders and to answer any questions they may have concerning the system, its specifications, or the bid procedures before bids are received.

Privately funded projects are far less restrictive than public projects because private money is funding the project, and, in essence, private money buys what private money wants. It is not uncommon to find that a corporate client may specifically restrict bidding to one or two controls manufacturers who have worked with that client in the past and have proven their capabilities. This situation in many respects lessens the liability of the design consultant and eases the process of specification preparation. Concerns over the proprietary nature of the specification can be answered by the developer of the project, who has no problem in confirming a stated preference for one vendor over another.

Operator Training and Support

Documentation

Complete documentation of the system installed shall be provided to the owner upon completion of the project. This documentation shall include general instruction sheets on each piece of hardware in the system, a complete drawing or graphic of the entire control system including all mechanical devices under control, and a hard-copy printout of all program algorithms used in the programming of the DDC computer.

Properly prepared and maintained documentation will provide future building operators with a complete reference on the system and its operation and will help current building operating personnel locate and solve problems.

Software Backup

All program routines in the DDC computer should be backed up on floppy diskettes or on a more permanent medium, such as a Winchester disk, and should be kept both on-site and off-site to protect the system in the event of a programming failure or if the memory of the computer should be accidentally erased.

This backup will allow for an expedient programming download and system restart. Because minor changes are made to programming as the bugs are worked out of the mechanical system, it is important to maintain the integrity of the software backup. It is recommended that a backup of computer programs be taken at least once a month during the first year of building operation.

Operator Training

Building operations personnel are charged with the practical matter of operating and maintaining the HVAC systems in a building. Because building operators rarely have an effect on the design of the buildings they manage, it is important that the building operations personnel fully understand the control strategies applied to their mechanical systems and how the controls are supposed to operate under varying conditions. Therefore, a minimum requirement of operator training of at least 40 classroom hours is recommended on every DDC project. It is quite common to find that after a period of time has passed since building commissioning, changes in building operations personnel may occur and people may be in charge of the HVAC control system who have had no formal training in its operation and use. Therefore, it is recommended that the controls contractor provide an additional 24 hours of training 12 to 17 months after completion of the system installation. Even if the same building operations personnel are still in charge of the system, they will benefit from a refresher course in how to obtain the greatest use of the control system the building owner has invested in. Furthermore, over time it is not uncommon to find that building operators will not use many of the features and functions available to them in a DDC system. A refresher course on system application and use will renew interest in using the control system to its fullest extent.

Off-Site Telecommunication Support

It is recommended that every DDC system be provided with a modem to allow remote communication with the control system. This is one of the key benefits of microprocessor-based control systems. A building operator can call in to a control system from another job site or even from home, using a personal computer or portable laptop computer, and can quickly monitor the operation of an HVAC system or make minor changes to the attributes of control system. Another benefit is on-line support from the installing contractor. If a building operator should run into trouble, he or she can contact the installing contractor, who can then dial the control system by using a personal computer and diagnose and correct problems without dispatching a service technician to the job site. This lowers warranty costs for the installing contractor and greatly reduces the downtime due to a system failure. Software has been developed, and is in use, that has the capability of dialing out of the DDC control system through the modem to alert a building operator or other designated person that a problem exists on the job site. For instance, in the event of an alarm the DDC will call up a book of telephone numbers and will sequentially call each number until an acknowledgment has been received from the receiver of the call by entering the appropriate pulse-tone digit from a touch-tone telephone. Some dial-out systems even have the capability of providing voice emulation; when you answer the telephone, a computer-generated voice explains the nature of the alarm condition and offers a series of actions to take by pressing buttons on a touch-tone telephone.

System Maintenance

It is recommended that, in addition to system warranty provided by the installing contractor, a separate price for system maintenance be provided at the time that the base system bids are submitted. The theory here is that the cost for such maintenance is most likely to be lower when it is being evaluated along with the price of the base control system. This maintenance program should include all hardware, software, and system wiring and should be based on a periodic maintenance program along with a minimum guaranteed response time to system failures. All of this should be provided over and above the standard warranty that the manufacturer and installing contractor provide, which is usually a period of 1 year from owner acceptance of the system.

It is also recommended that an extended warranty be provided by the installing contractor that covers every facet of the control system. In this manner, the building owner assured that the control system shall be properly installed and maintained and that minor problems will be repaired at no additional cost over a specified period of time in the future. A 2-year extension to the original 1-year warranty is recommended on large systems, rendering a minimum of 3 years of guaranteed support to the control system. The cost of such warranty and maintenance is minor when compared with the size of the investment that a DDC system represents. All too often a sophisticated control system erodes because of improper usage by building operators and lack of proper maintenance. It is strongly recommended that a comprehensive program of operator training and support, extended system warranty, and periodic guaranteed system maintenance be provided immediately following system installation and commissioning.

Additional Operation and Maintenance Considerations

The purpose of a good operation and maintenance program is to ensure that the desired level of system performance is maintained and to provide the operator with confidence that the data being received from the DDC system are accurate and that all control strategies are being executed properly. It is important to recognize the elements of a maintenance program for a DDC system.

Many people mistakenly believe that a computer requires no maintenance. This is simply not the case. The maintenance of a central computer is provided in the form of diagnostic programs. A diagnostic program indicates and identifies the failure of any component in the system, including a failure of the software. Such a failure is indicated to the building operator through a failure alarm. Many times the repair of a failed microprocessor-based component requires nothing more than a new chip or circuit board, resulting in little downtime. It is important, however, to have an ample supply of replacement parts either on site or as part of the inventory of the installing contractor. Nothing is more frustrating than having to bypass a control loop or an entire field control panel because a replacement part is on back order. Therefore, it is recommended that a minimum stock of replacement parts be on hand at all times, and this consideration should be weighed heavily in the evaluation of an installing contractor.

Also important to a proper maintenance program is the maintenance of HVAC equipment. A DDC system can provide service and maintenance reminders as well as printed maintenance schedules for HVAC equipment that can be triggered by operation time or calendar days. Operation time, also known as run time totalization, alerts the building maintenance personnel to the need for maintenance. Maintenance time reminders should be acknowledged, indicating to the DDC system that the maintenance has been performed: Ideally, the DDC system should be designed to permit an operator to acknowledge maintenance reminders only after such services have actually been performed. Control over the acknowledgment functions on the computer should be limited to management-level personnel only and should be protected from unauthorized acknowledgments by passwords. The last part of an effective operation and maintenance program is maintaining accurate calibration of sensing devices.

The outputs of many local-loop sensors and transmitters will change to some degree over time. There are few combinations of sensors and transmitters that can be expected to remain in tight calibration longer than 6 months to a year, and this is especially true for pneumatic devices, which are known for their propensity to drift. RTD sensors rarely drift over time, which is why they are so desirable in DDC systems. However, the characteristics of their installation may subject them to insulating materials, such as dust. Therefore, it is important to make sure that they are periodically cleaned and checked for damage.

Experience has shown that field devices, if they fail, are most likely to fail shortly after installation owing to either an installation error or a manufacturer's defect. Many times the failure of one device in an entire system goes unnoticed; however, the effect of such a failure has a ripple effect throughout the control system. An effective diagnostics program provided in the host computer should be capable of locating the source of the error and providing an indication to the building operator as to the nature of the failure and where it is physically located in the system.

System Commissioning

System commissioning includes the checkout, testing, and verification of system operation in every detail and at every level of the system. The procedures and operations of a system commission must be described in the project specifications in detail together with the minimum performance levels that must be established.

The commissioning of a DDC control system must be performed in a systematic and methodic manner and usually begins at the most distributed level of the system, which are the sensors and control devices. The actual field installation of sensors, actuators, and control devices must be checked. The integrity of the wiring to each device must be verified, and the integrity of the control signals over the wiring must be confirmed. The next step is to calibrate each sensor and to stroke each actuator manually to confirm that it has been properly linked. The values of every sensor and the position of each actuator must be verified at the DDC panel to ensure that the field computer is properly reading the information that is coming to it. This process is known as a *point test.*

After all points in the system have been verified for proper installation and operation, the next step is to perform a local-loop test. In a local-loop test the sensors and control devices that made up individual loops must be tested and observed for the proper response under full-load and partial-load operating conditions. For example, a sensor in the supply-air duct measuring the supply-air temperature is responsible for the modulation of a two-way valve on a cooling coil. This is one loop. After an artificial load is introduced on the sensor, the valve modulation is then observed to verify that the loop has been properly tuned and operates correctly.

Once all local loops have been tested and proven, the next step is to test the supervisory functions of the system. Time clock functions and specific supervisory control strategies should be checked for each mode of system operation. For example, at 7:00 a.m. the desired building temperature might be 72°F, the supply-air temperature may be set at 55°F, and the speed of the supply fan may be 100%. Each time-based sequence of control should be checked by manually resetting the time schedule of the DDC control system to observe the proper operation of the system.

The process of point tests, local-loop tests, and supervisory function tests must be performed for each DDC field panel in a system. Once all of the field DDC panels have been verified for proper operation, the checkout of the host computer system can begin. Once again, end-to-end sensor and actuator verification is necessary to make sure that the host computer database is reading the correct point identification for each device and is receiving the proper information.

After the integrity of the incoming data has been verified, graphics can be developed for each system on the local area network and testing of the host software can begin. Dynamic point information that is incorporated into graphic displays should be verified against actual field conditions. For instance, the value being read by a temperature sensor should be manually checked against the value of that sensor as shown on the graphic display. This verification should be done at the field level, at the field control panel level, and at the host level for each point on the system. Once this has been done, the database and trending functions of the host computer can be checked. Sample reports of all points on the system should be printed out and compared to data at the field control panel level and at the field device level. Finally, a thorough test of the entire system should be completed by randomly changing system attributes and observing the behavior of the system in response to such changes.

Specify to Avoid Hazards

There are four primary reasons why specifying engineers, contractors, and building owners run into problems with DDC systems:

- Poor design
- Poor installation
- Unreliable vendors
- Poor software

Poor Design

Poor design results from a lack of product and application knowledge on the part of the consulting engineer and, unfortunately, is all too common. The application of DDC products requires a solid understanding of control systems engineering and a fundamental knowledge of computers and microprocessor-based controls. The many benefits of proper system design have been established in this chapter.

Poor Installation

Poor installation results from the installing contractor's lack of knowledge. Common problems include running control signal wire in the same raceways as power wiring, thereby polluting the control signal. Other common problems in DDC system installation include improper grounding of shielded conductors. Great care must be taken in the installation of a DDC system to maintain the integrity of the control signal paths between the devices in the system. Most of the problems encountered in the field on a DDC system involve wiring.

Unreliable Vendors

Unproven and untested products present a great risk to the system designer and building operator. Once a system is installed it is very expensive to replace in terms of replacement cost and downtime associated with the installation of replacement equipment. Therefore, it is important to evaluate, preselect, and specify acceptable products during the design process. The manufacturer qualification and preselection process is highly recommended on any job of significant size. On smaller DDC projects, it is perfectly acceptable to specify three or four products to the exclusion of all others. Although some manufacturers may feel that such a practice is discriminatory, it is in the best interest of the building owner.

Poor Software

A properly designed and installed DDC system can be rendered virtually useless if the software that is supposed to operate the system does not work properly. Errors in the program code, or "bugs," can cause inconsistent operation, loss of important trend data, system downtime, and even catastrophic "crashes," which is the violent shutdown of the system caused by a software failure. It is recommended that during the product evaluation stage a complete evaluation and analysis be made of the software packages proposed by each manufacturer to be certain that the software is tested, proven reliable, and has been used successfully in projects of similar size and scope. Even when fully functioning software is utilized, a new building usually requires a debugging period of at least 6 months to a year to get the building operating efficiently. Therefore, one can imagine the length of time it would take and the amount of effort required to solve control problems with inadequate software.

Chapter 9

Economic Analysis of Direct Digital Control Systems

In Chapter 7, a methodology for designing direct digital control (DDC) systems was suggested to help organize the control design process. The point was made that design decisions are affected by economics—the first cost of the system versus the benefits that it provides for the purpose of reaching a balance between function and affordability.

DDC was developed to make HVAC systems more energy efficient. If such efficiencies were not possible, only simple two-position control systems would exist today. However, the degree to which sophisticated controls are applied must be justified by operating efficiencies. The cost of a simple two-position control system is far less than that of a DDC system, except when consideration is given to the differences in system energy consumption between the two. The incremental energy saving produced by intelligent control systems must more than offset the incremental cost premium over conventional controls. The economic justification process must be applied to each control system to ensure that system selection and control application are in balance.

The first part of this chapter focuses on estimating methods used to calculate HVAC system operating costs and energy savings. Evaluations of specific control strategies are considered to prevent too many dollars from chasing too few BTUs or kilowatts. These calculations are factors in payback estimates used to weigh the feasibility of a control design. The second part of the chapter addresses control system cost estimating methods used during the preliminary, design, development, and construction bidding phases of a project.

Economic Analysis of Energy Saving Strategies

The three most effective ways to achieve energy saving in HVAC systems through the use of intelligent control systems are:

1. Reducing the speed of electric motors
2. Staging the operating time of high-energy-consuming equipment
3. Adjusting air and water temperatures

From these broad categories, both simple and complex control strategies can be created to reduce electric energy consumption. The first step in evaluating the effectiveness of an energy management system is to locate the areas of greatest potential energy saving in a system. It is known that lighting loads are the most demanding in terms of energy consumed; however, we will limit our discussion to energy strategies as applied to HVAC systems.

Table 9-1 summarizes, in order of priority, those energy saving strat-

Table 9-1 Summary of Energy Saving Strategies

Simple Strategies
1. Reduce fan and pump motor speeds using variable-frequency drives.
2. Stage the operation of chillers in multiple chiller systems.
3. Stage the operation of chilled- and hot-water pumps.
4. Stage the operation of cooling tower fans and recirculation pumps; modulate the speed of cooling tower fans based on condensor water return temperature.
5. Reduce demand on mechanical cooling and heating equipment using outside-air temperature economization techniques.
Complex Strategies
6. Reset cooling supply-air temperature to zones based on building demand.
7. (Dual-duct systems) Reset heating supply-air temperature to zones based on building demand.
8. (Dual-duct systems) Reset hot-water supply temperature to heating coils to modulate heating supply-air temperature in response to building demand; modulate the hot-water supply temperature until the hot deck maintains its set point with heating valves full open.
9. Reset chilled water supply temperature to cooling coils to modulate cooling supply-air temperature in response to building demand; modulate the chilled-water supply temperature until the supply-air temperature reaches set point with cooling valves full open.

egies where the greatest potential exists to curtail energy consumption. These strategies can be simple or complex, depending on how aggressive the energy saving strategy is. Simple strategies attack areas where large opportunities for energy saving exist; complex strategies involve the pursuit of smaller saving through the use of intelligent control systems. Although complex control strategies are possible through the high speed and accuracy of digital controllers, in practice the saving these strategies produce is sometimes offset by losses in other areas. For example, resetting the supply-air temperature of an HVAC system may increase the load on the pumping system serving its chilled- and hot-water coils. The net result is an increase in energy cost, which defeats the purpose of reset control as an energy conservation strategy. The design engineer must closely evaluate each control strategy under consideration to make sure that net losses do not occur. Once attractive energy reduction opportunities have been identified, estimates of power consumption must be made to determine whether a control strategy is economically feasible. Three of the most cost-effective strategies are discussed next.

Fan Energy Saving

Variable-air-volume (VAV) systems rarely operate at peak load. In fact, most VAV systems operate at partial loads for a large percentage of their operating hours. By modulating the speed of the supply fan to provide only enough air as is required to satisfy the heating or cooling load of the building, one obtains a substantial opportunity for energy saving. While several methods of reducing fan volume are available, the variable-frequency-drive method is the most efficient under partial-load conditions. This is because variable-frequency drives can maintain the efficiency of electric motors at very high percentages while operating the electric motor at very low speeds. Table 9-2 compares the kilowatt hour consumption of a 100-horsepower fan motor running at full speed to the same motor running at 80% of its rated speed over a period of 1 year. In this application, we are assuming that the fan will operate 10 hours a day, 5 days a week, 52 weeks per year for a total of 2600 operating hours. We are also assuming an average cost per kilowatt hour of $0.09. The annual kilowatt-hour savings achieved by reducing the fan speed an average of 20% over the year produces a 38% reduction of energy consumption. Table 9-3 summarizes the saving potential for the same motor if the overall motor speed were reduced even further. Moreover, by developing a load profile of a building based on annual weather data, we can develop a duty cycle chart for each building to provide an approximation of the fan speed and run time over the course of a year. With these data,

Table 9-2 Calculations of Variable-Speed Drive Savings (Constant Volume vs. Reduced Volume)

Assumptions	
Motor Horsepower	100
Hours of operation	2600
Cost per kwh	$0.09

Operating Cost at Constant Volume (100% Fan Speed)

% SPEED	% Hp	SYS. EFF.	% HRS	HRS/YR	KWH/YR	$/YR
100	90	0.9	100	2,600	193,882	$17,449.38
Annual Totals					193,882	417,449.38

Operating Cost with Variable-Speed Drive

% SPEED	% Hp	SYS. EFF.	% HRS	HRS/YR	KWH/YR	$/YR
100	100	0.9	0.00%	0	0	0
90	72.9	0.84	0.00%	0	0	0
80	51.2	0.83	100.00%	2600	119,599	10,764
70	34.3	0.82	0.00%	0	0	0
60	21.6	0.81	0.00%	0	0	0
50	12.5	0.78	0.00%	0	0	0
40	6.4	0.68	0.00%	0	0	0
30	2.7	0.54	0.00%	0	0	0
Annual totals			100.00%	2600	119,599	$10,764.00
Total annual kwh savings						74,283
Total annual $ savings						$6,685.38

very accurate estimates of energy cost saving can be developed. In Appendix I, a template for an electronic spreadsheet is presented that can be used to calculate the estimated operating cost saving using a variable-speed drive. By distributing the total hours of operation per year over the average fan speed percentages, we can calculate an annual total that is weighted to reflect the building load profile.

The investment payback in motor speed reduction strategies is usually quite attractive. For the example of the 100-horsepower motor in Table 9-3, at an average speed of 80%, the estimated cost saving is $6685. According to Table 9-4, for the average cost per horsepower to install a motor speed controller, the approximate payback for this application is approximately 5 years.

Table 9-3 Potential Operating Cost Savings by Reducing Fan Speed

Assumptions			
Motor horsepower		100	
Annual operating hours		2600	
Cost per kwh		$0.09	
Average Annual Fan Speed	**Annual KWH Consumption**	**Annual Operating Cost**	**Operating Cost Savings Potential**
100	215,424	$19,388	0%
90	168,262	15,144	22%
80	119,599	10,764	44%
70	81,099	7299	62%
60	51,702	4653	76%

Table 9-4 Approximate Cost Per Horsepower Variable Frequency Drives

Motor Horsepower	VFD Cost	Installation Cost	Installed Cost Per HP
250	$42,000	$2000	$176
200	32,000	2000	170
150	28,000	2000	200
100	21,000	2000	230
75	16,000	1600	235
50	12,000	1600	272
25	7000	1200	328
10	4000	1200	520
5	3500	1200	940

VFD cost data based on variable torque, 480-volt, 3-phase PWM inverter housed in a Nema-1 enclosure.

Source: Saftronics, Inc.

Chiller Staging

In multiple-chiller configurations, such as those described in Chapter 6, the start times of chillers should be staggered to reduce in-rush current. This can be easily done by using time schedule programs in a DDC system. Another way to reduce chiller operating expense is to sequence the staging of chillers based on condensor water-supply temperature. As variations in the building cooling demand occur, the quantity of chilled water required to maintain the cooling capacity of the HVAC system will change. Such fluctuations in building cooling demand make it difficult to pinpoint the actual hours of chiller operation required to satisfy load conditions. However, for every hour that a chiller is not operating, a significant energy saving is achieved. Chiller manufacturers can provide accurate estimates of the quantity of energy consumed per hour of operation.

Adjusting Temperatures

Several viable temperature reset strategies were discussed in Chapter 6 for air and water temperatures. When actual building load conditions are accurately measured by the DDC system, the temperature of supply air and water serving the system can be modulated to approximate the demand of the occupied space. These strategies are generally considered to be quite complex, and the potential for creating offsetting energy losses is high. Furthermore, such strategies require that all controls be integrated into the DDC system so that as much information as possible is available.

There are no benchmark values for the quantity of energy that can be saved by using reset strategies; each HVAC system must be evaluated individually for its potential. The determination of estimated energy saving for these methods can be determined only after carefully evaluating the equipment, its arrangement, and its operating conditions. It is beyond the scope of this book to evaluate calculations of this type; however, references are given in the bibliography for those who wish to pursue this area of economic analysis.

Estimating Installed Cost of Direct Digital Control Systems

Before we discuss estimating procedures for control systems, we review some common terms used in construction estimating. The practice of calculating the material and labor costs for a construction project is known as a *takeoff*. Separate material and labor takeoffs are made to provide the construction estimator with a firm understanding of all of the hardware required for a control system and the time that it will take to have it properly installed. Before any hardware is installed on a construction project, the conduit, wire, and tubes used to interconnect the control devices must first be installed. This phase is known as *rough-in*. Roughing in a job means that the interconnecting media has been installed but not connected to the devices it serves. The term *connection* refers to the actual termination of wires or tubes to control devices in the field. *Checkout* is the process of verifying the integrity of the wiring or tubing terminations and making sure that the interconnecting media provide an uninterrupted signal path to and from each control device. In summary, it is important to recognize that five separate scopes of responsibility exist whenever a control system is to be installed. The controls estimator must ask the following questions:

1. Who is going to furnish the control system hardware?
2. Who is going to install the hardware?
3. Who is going to rough in the interconnecting media?
4. Who is going to terminate all of the connections from the interconnecting media to the control devices?
5. Who is going to check out and verify the integrity of the system?

By answering these five questions, the estimator will be sure not to overlook any part of the control system estimate.

Estimating is as much of an art as it is a science. Understanding all aspects of a control system is necessary before the estimating process takes place. You must understand the scope of a building project to be able to identify and estimate for construction risks. Risk assessment and risk management are the hallmarks of a professional control contractor.

Budgetary Estimating Methodology

There are two primary ways to perform budgetary cost estimates: the point-cost method and the square-foot method. The *point-cost method* is a pricing model that is frequently used at the predesign stage of a project. Often a consulting engineer or design-build contractor will want to know if a concept for a building control system is feasible from an economic standpoint. By comparing the costs of a per-point basis to those for similar buildings, you can derive a rough estimate of what a system should cost. You must be careful when using the point-cost method, however; there are many factors that can affect the accuracy of the cost per system point. First, point costs refer to physical points only. Consideration is not given to pseudopoints, because they exist in software only. Second, the point-cost method assumes that all building control systems are configured basically the same; this is rarely true. Office buildings, industrial buildings, and special-use buildings, such as hospitals, schools, hotels, and nursing homes, all use different control system architectures owing to the difference in building design. Frequently, buildings that have a similar purpose may have drastically different control designs because of the needs of the tenant. Furthermore, if a host or central computer is included in the job, this can raise the point-cost range. Table 9-5 provides point-cost range estimates for DDC that are categorized by control function. These point-cost ranges will vary from market to market and should be checked against market conditions in major metropolitan areas.

Table 9-5 Per Point Cost Comparisons: DDC Systems

Point Type	Hospital	Office Building	School/ University
Temperature input	$250	$250	$250
Flow (volume) input	375	375	375
Pressure input	350	350	350
Position control output	425	425	425
Speed control output	280	280	280
Start/stop control output	160	160	160
Flow (proof) input	160	160	160
Alarm indication input	185	185	185
Average total cost per point, installed	900	800	850

Values shown are approximate, and are taken from actual projects during 1988-1990.

The square-foot method is also a useful way to prepare a budgetary estimate on a control system. With the square-foot method, care must be taken to recognize the differences between control systems in buildings of various design and complexity. The calculation of the cost per square foot is normally derived by using the net leasable area as the denominator in the equation. Since the cost per square foot varies so greatly from building to building depending on architecture and the layout of the floor plan, an accurate table of values is not possible. Local control contractors can provide these data from recent projects that are similar to those under design.

There are several reliable sources of information on control system costs that are available in most major metropolitan areas. Control system manufacturers and control contractors are a good source of information on material, labor, and total installed cost. Of course, the accuracy of such budget estimates is based on that contractor's perception of his or her chances of winning the job. The greater the opportunity a contractor recognizes, the more accurate the estimate.

Lastly, it is helpful to review the installed cost of control systems for similar jobs. Again, controls contractors can be a good resource for this kind of information.

Hard-Cost Estimating Methodology

Once a control system design has reached the final stages of design, it can be submitted to contractors for a firm bid. The following seven-step methodology is suggested for estimating the hard cost of control systems:

1. Identify all control devices in the system.
2. Determine the physical location of control devices on the floor plan.
3. Determine the type and size of all interconnecting media (wire, conduit, and tube).
4. Determine material cost and installation labor hours.
5. Adjust the cost estimate for construction conditions.
6. Apply contractor's markup.
7. Adjust estimated price for competitive conditions.

Material Takeoff

The first step in the estimating process is to determine the quantity and type of materials required for the control system. Material estimating is the most accurate area of control system estimating. A well-designed system, presented on a well-documented set of plans and specifications, will tell the estimator the components and operating functions required of the system.

The simplest and most organized way to estimate material cost is to group materials by unit, type, and loop. For instance, the controls operating an air-handling machine should be segregated in the estimate by the plan designation of the air handler (i.e., AH-1, AH-2, etc.). Because the control of an air handler involves several separate control strategies, and because strategies vary with the size of an air handler, it is most convenient to estimate by control loop. For example, in large distributed control systems where simultaneous loop control of HVAC equipment is being done, it is best to separate control components by loop functions, such as mixed-air control, supply-air control, or zone control. Examples of both organization methods are shown in Figure 9-1.

Once all materials have been grouped, the unit cost for each device can be applied to each line item and the columnar data can be totaled. There are several good sources for control component prices. Control system contractors, control component wholesalers, and control system manufacturers are all reliable sources for accurate material cost data.

Interconnecting Media

The next step in the material estimating process is to determine the type and length of interconnecting media required to tie the control system together. Again, a sensible approach to estimating interconnecting media is to organize these data by groups. For example, sensor wiring, local area network (LAN) wiring, pneumatic tubing for thermostats, and main-air tubing to serve controllers on pneumatic VAV boxed are all separate categories of media that should be estimated

Control Item/Task Description	Unit Price	Unit Labor Rate	Material Price	Labor #1 Hours	Labor #2 Hours	Labor #3 Hours
AH-1 Control:						
MA Loop:						
Sensors						
Controlled Devices						
SA Loop:						
Supply Air Sensor						
Room Sensors						
etc.						
		Totals				

Figure 9-1 Material and labor estimate takeoff sheet.

separately. This is a simple process of measuring the distance from one device to another from the floor plans.

When you are doing a media takeoff, keep in mind that a blueprint is a two-dimensional depiction of a building and does not necessarily reflect actual field conditions. Calculations of horizontal and vertical distances must be made and distance calculations must be adjusted to allow for restrictions and obstructions when the media are actually being installed. As a rule, it is best to make all measurements at right angles. Rough-in work in mechanical equipment areas is especially restrictive owing to limited space; all right-angle distances taken from mechanical areas should be increased from 30 to 50% to accommodate tight working conditions that may require the use of additional conduit or tubing to complete the installation. Another important consideration in rough-in takeoff is ceiling heights. It is easy to forget how high a ceiling is when one is looking at a two-dimensional blueprint; be sure to check ceiling heights and also notice the location of beams and columns, and take into consideration their impact on the distance of interconnecting media. Figure 9-2 provides an [illegible]ized method of arranging rough-in calculations so that the appropriate factors can be applied.

Total all distances by media type and apply material costs to each. The material cost for rough-in materials can be obtained from electrical

Conduit/Tube, Type/Size	Distance	Unit Material Price	Material Factor	Total Material Cost	Unit Labor Rate	Labor Factor	Total Labor Hours
		Total Material Cost:			Total Labor Hrs:		

Figure 9-2 Rough-in labor and material summary.

wholesale supply houses, as well as plumbing supply stores for pneumatic tubing. These materials are usually quoted on a lineal foot basis.

The quantity, type, and locations of control devices can be determined from the control system schematic drawings, and the locations of such devices in a building can be taken from the mechanical blueprints prepared for the project. The determination of the type and size of interconnecting media can be obtained from reviewing the manufacturer's data for the control devices in question. These data will provide guidance on the size of conductors and the minimum requirements for conduit and tubing to support the operation of such devices. Material costs can be gathered from various suppliers, including manufacturers, wholesalers, and regional representatives. The calculation of installation labor hours, which will be detailed here, will vary from region to region, depending on the labor rates for electrical and pipe-fitting work in an area. Union halls publish an hourly cost breakdown for the various types of labor and categories of workers who are approved to perform such labor.

Labor-Cost Estimating

The end goal of labor-cost estimating is to determine accurately the amount of time that it will take to mount a control device and terminate power and signal sources. Labor estimating is the riskiest area of construction estimating because there are no guarantees as to how long it will take to perform a given task. For purposes of organization, the installation of control system devices is broken into two succinct categories: component (or hookup) labor and media labor.

Component labor involves the time that it will take to install, connect and terminate, calibrate, and final test each device in the system. It is estimated by using a labor factor on a per unit basis. Figure 9-3 summarizes the hookup labor factors for common types of control devices. Each factor is quoted in units of labor hours. For example, the time that it takes to hook up a single damper actuator is 1.5 hours. The cost of labor is derived by calculating unit labor cost against the total quantity of hookup labor. After hookup labor has been estimated, an allowance must be made for the time that it will take to terminate, test, and calibrate each device. This factor is also quoted on a per unit basis just like hookup labor and, in many cases, both hookup labor and termination time are quoted as a single figure in some published estimating manuals. In Table 9-2 these values have been broken down for your evaluation.

The next step in estimating the labor cost for a control system is to determine the amount of time that it will take to install the interconnecting media. Labor factors for media are quoted on a per-foot basis

Total Hours Labor: (Cat. #1)	____ x $ ____	/hr. =	$ ____________
Total Hours Labor: (Cat. #1)	____ x $ ____	/hr. =	$ ____________
Total Hours Labor: (Cat. #1)	____ x $ ____	/hr. =	$ ____________
Total Control Material Price			$ ____________
Total Installation Material Price			$ ____________
Programming/Training Allocation			$ ____________
Engineering/System Startup Allocation			$ ____________
Other Costs			$ ____________
		Subtotal	$ ____________
	Overhead Rate ______	Overhead	$ ____________
		Total Cost	$ ____________
	Profit Rate ______	Profit	$ ____________
		Material Sales Tax	$ ____________
		Total Price	$ ____________

Figure 9-3 Control estimate recap.

and vary in relation to the size and type of media that must be installed. For example, fire-rated tubing, which is flexible and fairly easy to maneuver, will take far less time to install on a per-foot basis than a rigid metal conduit, which can be difficult to work with because it is heavy, inflexible, and requires threading to install. Figure 9-2 summarizes the rough-in labor factors on a lineal foot basis for several types of common interconnecting media. After the total media labor hours required for rough-in are totaled, they must be adjusted by applying unit labor factors that address difficulties in installation. Physical obstructions to installation, as well as other project constraints, must be taken into consideration. Multipliers are commonly used, depending on the overall size of a job as measured in total rough-in hours. Refer again to Figure 9-2. Common rough-in labor multipliers are used for difficult ceilings, crowded mechanical rooms and shafts, interconnecting media systems requiring wiremold or work within occupied remodel spaces, and additional labor that must be accounted for when trenching or conduits between buildings are required. The best approach to take at this phase of an estimate is to visualize the system as it will appear when it is being installed. By visualizing the constraints to installation and by evaluating each risk factor, you can understand how working conditions and/or code restrictions can tremendously affect the progress and overall outcome of a wiring installation.

Once the hookup labor and rough-in labor hours have been estimated and summarized, they should be summarized on an estimating recap sheet, as shown in Figure 9-3. On a single spreadsheet, the description of each device, quantity of each, cost per device, hookup labor, and rough-in labor can be evaluated on a per-unit basis. This takes care of the physical aspects of furnishing and installing a control system. Once the rough-in work and component installation have been completed, the next step is to install the software and control program algorithms required that will instruct the system how to operate.

Programming Labor

The time required to program an automation system correctly is a function of the number of physical points on the system as well as the pseudopoints that exist on the system. For example, say you have a system with 200 physical points and 40 pseudopoints; the 40 pseudopoints would be points that provide information to the building operator based on the activity of the 200 physical points. The total quantity of points on the system requiring programming is 240. The time that it takes to program each point is between 1 and 3 hours, depending on how "friendly" and flexible the programming software is.

In addition to point programming time, you must also estimate the time it will take to build custom graphics. Graphic screens are generated by using on-screen tools provided with most commercially available graphic software packages. These packages also include standardized symbol libraries so that common shapes and forms, such as the standard shape for an air handler or for a water valve, are included. On average, it takes between 1 and 2 hours to generate a complete graphic screen. This figure also includes the time needed to locate real-time data fields and to test the screen to make sure that the data will populate the screen correctly. The last engineering labor component to estimate for is the time that it will take to prepare adequate documentation of the control programs, graphics, and overall operation of the system. This value will vary in proportion to the size of a job and can range from a minimum of 8 hours to a maximum of 2 weeks.

Estimate Recap

Once all cost data for material and labor have been collected and summarized, consideration must be given to the overall nature of the project and the potential for problems that the contractor might face in the performance of the contract. This is the subjective part of estimating, because the burden is on the estimator to make an intelligent guess as to how much additional material and labor must be added to the contract price to allow for contingencies.

The amount of a contractor's markup will vary in relationship to the size of the bidding contractor's firm, its backlog of work in progress, current economic conditions, and the competitive climate that exists when the project goes to bid.

The quality of the control system design and the care with which the plans and specifications were prepared also have a dramatic impact at this phase of a project. A well-documented set of plans and specifications will encourage bidding and will lower the overall contractor markup on a project because he or she is likely to perceive less risk in a well-designed project than that of a project that has been poorly prepared.

As a rule of thumb in estimating, it is safe to assume that the total material and labor costs will be marked up 35% to 50% to cover the general and administrative costs of owning and operating a construction company and allowing for a reasonable profit. This percentage is not a gross margin percentage, but is a markup or mark on percentage. For example, if material and labor total $1000.00, the contractor will probably bid $1350.00 to $1500.00.

Energy management and control systems have proven effective in reducing the total quantity of energy consumed in the operation of HVAC systems. It is quite common to find that a saving of 10% to 20% of total energy usage is possible with a well-designed and -installed EMS. The magnitude of energy saving will vary, depending on building type and hours of operation; obviously, a 24-hour facility has little energy saving opportunity, whereas a college campus, which is occupied intermittently, has a tremendous potential for energy saving.

The key to yielding the greatest saving from a control system is to combine energy conservation strategies and equipment technologies in the most efficient way possible. Because digital systems can control pneumatic and electric devices, and because they can be programmed to perform site-specific control routines, opportunities for energy reduction can be implemented in a cost-effective manner.

Chapter 10

Emerging Technologies in Direct Digital Control

The continuing evolution of microprocessor design and manufacture has fostered a similar evolution in the development of direct digital controllers. As large-scale integration and lower cost have made microprocessor power accessible and affordable to manufacturers of all kinds of devices, the usage of microprocessor intelligence has been driven down to the lowest possible levels of modern control systems. In short, intelligence is everywhere. What remains lacking in the digital control systems of today is the ability to connect all of this intelligence in a way that is meaningful and valuable to building operators.

Nonhierarchical Systems

Figure 10.1 depicts the evolution of DDC system topologies from 1970 to the present. If we can rely on the past as a guide, then future incarnations of direct digital control (DDC) will be even smaller and faster, using standard interconnecting media that are inexpensive and that will transmit information in a language common to all devices on a network. Within a matter of just a few years, these systems will render as clunky, proprietary, expensive, and difficult the technologies we use today.

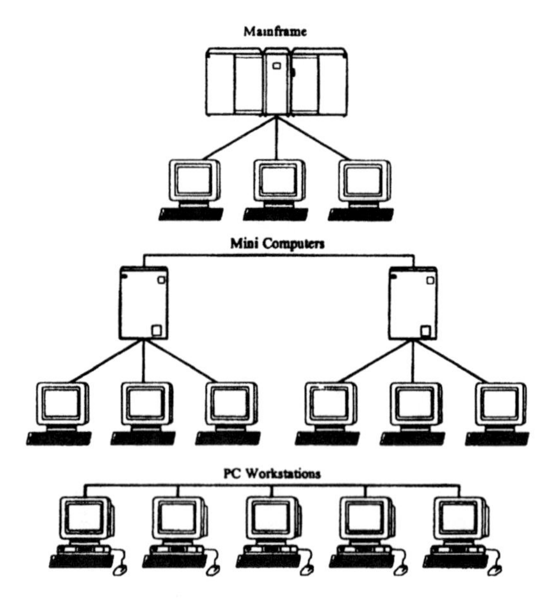

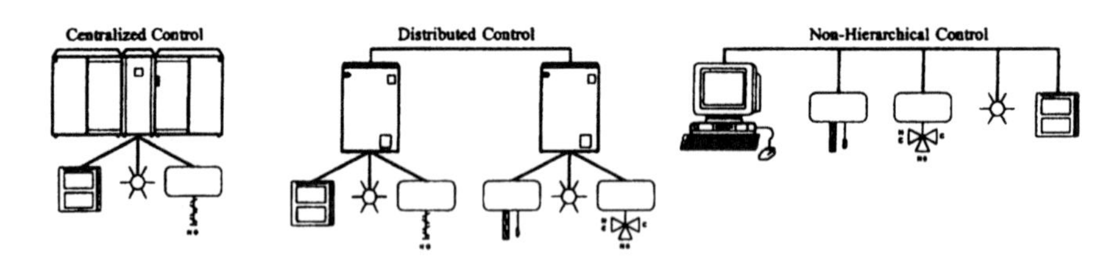

Figure 10-1 The Evolution of DDC Systems 1970–1997

In Chapter 5, both a vertical (BACnet) and a horizontal (Echelon) approach to expanding system communicability and interoperability was examined. Looking forward, we can expect to see continued innovation in the area of communication technologies for DDC systems.

Pattern Recognition Adaptive Control ("PRAC")

At the controller level, one of the more interesting developments in control theory is known as *pattern recognition adaptive control*, or *PRAC*. This is a scheme for automatically adjusting the proportional and integral values of a common proportional plus integral (PI) control loop based on patterns the DDC controller recognizes in the behavior of the control loops it is managing. Although the science underlying this control scheme has been in existence for some time, it is only recently that DDC manufacturers have begun to use this technology.

With the continued advancements in DDC system technology has come a certain complexity and challenge in how to use this equipment to its maximum potential. Today's DDC control systems are often time consuming to program and properly tune and in many cases systems are commissioned without these tuning procedures being fully completed. This results in control system operations that needlessly cycle equipment, causing wasted energy and physical wear on the equipment being controlled. This in turn leads to building comfort problems and unfortunately gives building owners a poor impression of the true capability of DDC systems. From these circumstances came the need for a new approach to conventional, PI control, namely PI systems that could automatically self-tune themselves based on observed patterns of behavior in control loops over time. This would mean that systems that were either incorrectly set up at installation or not fully completed in terms of their PI loop programming could self-correct once the system was in operation. Moreover, adjustments for seasonal changes and other sudden changes in load conditions could be automatically made without the need for retuning the entire DDC system. In Chapter 3, proportional plus integral plus derivative (PID) control was defined and explained. Most DDC controllers today are equipped to provide PID control; however, the derivative function (D) is rarely used as it has been proven to cause overreactions in sensitive HVAC system operations. You will recall that the derivative function manages the rate of change of the control input and because HVAC systems are by their very nature slow acting systems, sudden responses to changes in the control variable that is managed by the derivative function can cause the controller to throw the system into chaos. Therefore, it is a rarely used function owing to the instability it can create. This leaves us with PI control, which is the most common method of controlling HVAC control loops. The tuning parameters of a PI control loop are the *proportional band*, the *integral band*, and the *dead band*, the most important being the proportional band. The proportional band is the band that determines the stability of the control loop. PRAC optimizes the proportional band of the PI control scheme and makes adjustments that will finally tune the operation of this control loop to manage a given set-point.

To do this, the PRAC function monitors the controlled variable (temperature, humidity, pressure, etc.), the system set-point, and the output of the controller (a valve actuator, for example). The PRAC function then intelligently adjusts the proportional band and the integral or time function of the controller to make this adjustment. These adjustments are made in real time as the system is in operation and there is no limit to the number of adjustments the PRAC function will make to smooth the operation of the control loop and to optimize its efficiency in managing the system set-point.

When a building operator or system operator changes a set-point, which is a very common occurrence, the PRAC function recognizes the set-point change as a separate event, measures the closed loop control response of the new set-point, and recalculates the optimal proportional band and optimal integral function based on the new set-point and its ability to control to that set-point.

In the event that there is a sudden disturbance to the controlled variable, the PRAC function is able to recalculate quickly an optimum proportional band and, if the controlled variable suddenly returns to its original level, can quickly move this optimal proportional value back. This means that the tuning parameters of the PI control loop can be automatically updated as the system that it controls remains in continuous operation.

To date, PRAC algorithms have successfully managed HVAC control systems in educational facilities, laboratories, retail and commercial environments, manufacturing environments, and in laboratories in health care facilities. Numerous tests of this technology have proven that it is capable of managing HVAC systems in dynamically changing environments to within 5% of ideal performance conditions in over 90% of the tests that were made. This ability to tune DDC loops automatically to their optimum performance results in cost savings to building owners both in terms of ongoing maintenance and in the areas of comfort and energy conservation.

In addition, because PRAC algorithms are so simple, the processing power in today's DDC controllers are more than adequate to handle these new functions. Therefore, we can expect to see more PRAC-based PI control schemes in digital controllers in the future.

Appendix **I**

Spreadsheet Template for Motor Energy Savings Using Variable-Frequency Drives

This Appendix contains a spreadsheet template that can be used to estimate the energy savings and operating cost savings that can be achieved by reducing motor speed on HVAC equipment such as fans and pumps. This spreadsheet was written on Excel® for the Macintosh® and the formulas and functions used can be easily adapted to other spreadsheet programs.

The assumptions of operating conditions can be changed to suit the specifications of a given motor. Also, the distribution of operating hours that the motor runs under a given speed can be altered to match the load profile of a motor in an actual application. This is done by entering different values, totaling 100%, in cells D:18 through D:25.

	A	B	C	D	E	F	G
1	CALCULATIONS OF VARIABLE–SPEED DRIVE SAVINGS						
2	Constant Volume vs. Variable Volume						
3							
4	ASSUMPTIONS:						
5	MOTOR HORSEPOWER		100				
6	HOURS OF OPERATION		2,600				
7	COST PER KWH		$0.09				
8							
9	OPERATING COST WITH CONSTANT VOLUME						
10	% SPEED	% HP	SYS. EFF.	% HRS	HRS/YR	KWH/YR	$/YR
11							
12	100	90	0.9	100	2,600	193,882	$15,510.56
13							
14				ANNUAL TOTALS		193,882	$15,510.56
15							
16	OPERATING COST WITH VARIABLE SPEED DRIVE						
17	% SPEED	% HP	SYS. EFF.	% HRS	HRS/YR	KWH/YR	$/YR
18	100	100	0.9	10.00%	260	21,667	1,733
19	90	72.9	0.84	15.00%	390	25,239	2,019
20	80	51.2	0.83	50.00%	1,300	59,800	4,784
21	70	34.3	0.82	25.00%	650	20,275	1,622
22	60	21.6	0.81	0.00%	0	0	0
23	50	12.5	0.78	0.00%	0	0	0
24	40	6.4	0.68	0.00%	0	0	0
25	30	2.7	0.54	0.00%	0	0	0
26			ANNUAL TOTALS	100.00%	2,600	126,981	$10,158.00
27							
28			TOTAL ANNUAL KWH SAVINGS				66,901
29			TOTAL ANNUAL $ SAVINGS				$5,352.56
30							
31							
32							
33							

	A	B	C	D	E	F	G
1	**VSD SAVINGS- FORMULAS**						
2	**Constant Volume vs. Variable Volume**						
3	**Fan Motor Application**						
4							
5	ASSUMPTIONS:						
6	MOTOR HORSE POWER		100				
7	HOURS OF OPERATION		2600				
8	COST PER KWH		0.1				
9							
10	OPERATING COST AT CONSTANT VOLUME						
11	% SPEED	% HP	SYS. EFF.	% HRS	HRS/YR	KWH/YR	$/YR
12							
13	100	=A13*C13	0.9	100	=C7	=$C$6*0.7457*B13/C13*E13/100	=$C$8*F13
14							
15				ANNUAL TOTALS		=F13	=G13
16							
17	OPERATING COST WITH VFD						
18	% SPEED	% HP	SYS. EFF.	% HRS	HRS/YR	KWH/YR	$/YR
19	100	=A19^3/10000	0.9	0.1	=$E$13*D19	=$C$6*0.75*B19*E19/C19/100	=F19*$C$8
20	90	=A20^3/10000	0.84	0.15	=$E$13*D20	=$C$6*0.7457*B20*E20/C20/100	=F20*$C$8
21	80	=A21^3/10000	0.83	0.5	=$E$13*D21	=$C$6*0.7457*B21*E21/C21/100	=F21*$C$8
22	70	=A22^3/10000	0.82	0.25	=$E$13*D22	=$C$6*0.7457*B22*E22/C22/100	=F22*$C$8
23	60	=A23^3/10000	0.81	0	=$E$13*D23	=$C$6*0.7457*B23*E23/C23/100	=F23*$C$8
24	50	=A24^3/10000	0.78	0	=$E$13*D24	=$C$6*0.7457*B24*E24/C24/100	=F24*$C$8
25	40	=A25^3/10000	0.68	0	=$E$13*D25	=$C$6*0.7457*B25*E25/C25/100	=F25*$C$8
26	30	=A26^3/10000	0.54	0	=$E$13*D26	=$C$6*0.7457*B26*E26/C26/100	=F26*$C$8
27			ANNUAL TOTALS		=SUM(E19:E26)	=SUM(F19:F26)	=SUM(G19:G26)
28							
29			TOTAL ANNUAL KWH SAVINGS				=F13-F27
30			TOTAL ANNUAL $ SAVINGS				=G13-G27
31							
32							

Appendix II

Direct Digital Control Systems Manufacturers

As the result of the relatively recent industry-wide acceptance of direct digital control (DDC) systems, the designs of many modern HVAC control systems are now based on distributed digital architectures. Following is a list of companies that offer high-quality DDC systems.

There are many companies manufacturing products called "energy management systems," which range from simple clocks to mainframe-based building management systems. Exercise extreme caution when evaluating products that lack a strong track record of successful installations. The following companies manufacture mature products that are frequently recommended by experienced specifying engineers and are presented here to assist you in locating sources for product information.

ALERTON TECHNOLOGIES, INC.
6670 185th Avenue N.E.
Redmond, WA 98052
Telephone: (425) 869-8400
Facsimile: (425) 869-8445
URL: www.alerton.com

AMERICAN AUTO-MATRIX, INC.
One Technology Drive
Export, PA 15632-8903
Telephone: (412) 733-2000
Facsimile: (412) 327-6124
e-mail: 70740.435@compuserve.com
URL: www.auto-matrix.com

ANDOVER CONTROLS CORP.
300 Brickstone Sq.
Andover, MA 01810
Telephone: (508) 470-0555
Facsimile: (508) 470-0946
URL: www.andovercontrols.com

AUTOMATED LOGIC CORPORATION
1150 Roberts Blvd.
Kennesaw, GA 30144-3618
Telephone: (770) 429-3000
URL: www.automatedlogic.com

BARBER-COLMAN COMPANY
Division of Siebe PLC
Environmental Controls Division
1354 Clifford Avenue
Loves Park, IL 61132
Telephone: (815) 637-3000
Toll Free: (800) 232-4343
Facsimile: (815) 637-5341
URL: www.barber-colman.com

CONTROL PAK INTERNATIONAL
P.O. Box 127
Walled Lake, MI 48390
Telephone: (248) 960-8800

CONTROL SYSTEMS INTERNATIONAL, INC.
4210 Shawnee Mission Parkway Suite 200A
Fairway, KS 66205
Telephone: (913) 432-4442
Facsimile: (913) 432-0392
e-mail: mktg@csiks.com
URL: www.csiks.com

HONEYWELL INC.
Building Systems Division
Honeywell Plaza
Minneapolis, MN 55440
Telephone: (612) 951-1000
URL: www.honeywell.com

HSQ TECHNOLOGY
1435 Huntington Avenue
South San Francisco, CA 94080
Telephone: (415) 952-4310
Facsimile: (415) 952-7206

JOHNSON CONTROLS INC.
5757 N. Green Bay Avenue
P.O. Box 591
Milwaukee, WI 53201
Telephone: (414) 228-1200
URL: www.jci.com

LANDIS & STAEFA
Division of Siemens AG
1000 Deerfield Parkway
Buffalo Grove, IL 60089
Telephone: (847) 215-1000
URL: www.us.landisgyr.com

NOVAR CONTROLS CORPORATION
24 Brown Street
Barberton, OH 44203
Telephone: (330) 745-0074

Bibliography

Amborn, Randy, et al. "Decision Steps for Implementing a BACnet Interface Project." *ASHRAE Journal*, November 1995, 34–39.

ASHRAE Energy Professional Development Series Handbook. *DDC for HVAC Monitoring and Control.* Various authors, 1986.

ASHRAE *HVAC Systems and Applications Handbook.* 1987, 51-1-51-13.

Barber-Colman Company Automation Systems Dictionary. 1982.

Bartos, Frank J. "Fuzzy Logic Reaches Adulthood." *Control Engineering*, July 1996, 50–56.

Brooner, E.G. *The Local Area Network Book.* Indianapolis, IN: Howard W. Sams, 1984.

Bushby, Steven T., et al, "Standardizing EMCS Communication Protocols." *ASHRAE Journal*, January 1989, 33–36.

Diamond, Mitchell S., et al. "New Opportunities for Energy Management Systems." *Strategic Planning and Energy Management*, Spring 1988, 34–36.

Dowd, John B. "VAV Energy Conservation Systems: Technical paper, Barber-Colman Company, 1973.

Fischer, David M. "The Intelligent Building: Fact of Fiction?" *Strategic Planning and Energy Management*, Fall 1987, 15–29.

Gladstone, John, et al. *Mechanical Estimating Guidebook for Building Construction*, 5th ed. New York: AACE/McGraw-Hill, 1987.

Haberl, Jeffrey S., et al. "Diagnosing Building Operational Problems." *ASHRAE Journal*, June 1989, 20–30.

Hahn, Warren G. "Is DDC Inevitable?" *Heating Piping and Air Conditioning*, November 1988, 81–95.

Haines, Roger W. *Controls Systems for Heating, Ventilating and Air Conditioning*, 4th ed. New York: Van Nostrand Reinhold, 1987.

Hall, John e., et al. "Computer Software Invades the HVAC Market." *ASHRAE Journal*, July 1989, 32–44.

Herdeman, James R. "Open Communications Protocol for Interconnectability." *Energy Engineering*, Fall 1989, 60–65.

Kelly, Kevin. "New Rules For The New Economy: Twelve Independent Principles For Thriving In A Turbulent World" *Wired*, September 1997, 141–143.

Krutz, Ronald L. *Microprocessors for Managers*. Boston: CVI, 1983.

LeBlanc, Richard J. "Direct Digital Control . . . Is It Really Better?" MCC Powers Corporation, 1–7.

Lundstrom, C.E. "Direct Digital Controls Compare Favorably in HVAC Installations." *Energy Engineering*, November 1988, 17–34.

Madan, Pradip. "Overview of Control Networking Technology." *Echelon Corporation*, 1997.

Madron, Thomas W. *Local Area Networks*. New York: John Wiley & Sons, 1988.

Mayhew, Frank W. "Cooling Tower Control with Variable Speed Drives." Technical paper, *Energy Technology Conference*, 1987a.

Mayhew, Frank W. "Energy Savings Opportunities with Multiple Chiller Installations." Technical paper, *RETSIE/IPEC Conference*, 1987b.

Meredith, Jack. "Customized Control Strategies with High Performance EMCS." *Energy Engineering*, Volume 86, No. 2, 1989, 12–21.

Mixon, W.R. "Protocols to Guide Building Energy Monitoring Projects." *ASHRAE Journal*, June 1989, 38–42.

Norton, Peter. *Inside the IBM PC*. Santa Monica, CA: Brady Books, 1986.

Sandler, Corey. *How to Telecommunicate*. New York: Henry Holt, 1986.

Seem, John E. "A New Pattern Recognition Adaptive Controller" Technical paper, *International Federation of Automatic Control*, 1996.

Senior, Ken, et al. "Dinosaurs Alive in Mechanical Specifications." *Building Design & Construction*, November 1988, 67–69.

Trane Company. "A Need for Variable Flow Chilled Water Systems." *Trane Engineer's Newsletter*, Volume 6, No. 9, October-November 1977.

Trane Company. "Control of Water Chillers in a decoupler Variable Flow System. *Trane Engineer's Newsletter*, Volume 6, No. 10, January 1978.

Turner, Wayne C., Editor. *Energy Management Handbook*. New York: John Wiley & Sons, 1982, 347–350.

Wiegner, Kathleen K. "A Conflict of Interests." *Forbes*, November 1988, 251–255.

Williams, Verle A., et al. "Direct Digital Control: Benefits, Concerns and Pitfalls." *Strategic Planning and Energy Management*, Fall 1987, 4–14.

Index

Lightning Source UK Ltd.
Milton Keynes UK
18 September 2009

143892UK00001B/165/A